Non-Clinical
Vascular Infusion
Technology

Volume I

The Science

Non-Clinical Vascular Infusion Technology

Volume I

The Science

Owen P. Green and Guy Healing

CRC Press

Taylor & Francis Group

Boca Raton London New York

CRC Press is an imprint of the
Taylor & Francis Group, an **informa** business

CRC Press
Taylor & Francis Group
6000 Broken Sound Parkway NW, Suite 300
Boca Raton, FL 33487-2742

First issued in paperback 2018

ISBN-13: 978-1-4398-7449-3 (hbk)
ISBN-13: 978-1-138-38216-9 (pbk)

Library of Congress Cataloging-in-Publication Data

Green, Owen P., author.
  Non-clinical vascular infusion technology / Owen P. Green, Guy Healing.
  p. ; cm.
  Includes bibliographical references and index.
  ISBN 978-1-4398-7449-3 (hardcover : alk. paper)
  I. Healing, Guy, 1961- author. II. Title.
  [DNLM: 1. Drug Evaluation, Preclinical. 2. Infusions, Intravenous--methods. 3. Models, Animal. QV 771]

RM170
615'.6--dc23   2013027681

Visit the Taylor & Francis Web site at
http://www.taylorandfrancis.com

and the CRC Press Web site at
http://www.crcpress.com

Contents

Foreword

Good science and good welfare go hand in hand. Innovative science and technology can be used to improve animal welfare. Equally, the 3Rs (replacement, refinement and reduction of animals in research) can provide fresh insight and novel approaches to advance science. By sharing data, knowledge, and experience on the science behind infusion models and the refinement of the techniques used, there is the potential to have a significant impact on the 3Rs.

Owen Green and Guy Healing have shown the importance of the 3Rs in infusion technology at international meetings and in producing this book. The book will enhance uptake of the latest science behind vascular delivery systems to get better data and help influence decisions around the most appropriate model. It will also contribute to preventing repetition of method development and optimising experiments to answer specific scientific questions with the least impact on animals.

As in other areas of science, the field of infusion technology is constantly evolving. This edition of *Non-Clinical Vascular Infusion Technology* reviews current developments in the field that will support scientists in putting the 3Rs into practice.

Kathryn Chapman
Head of Innovation and Translation, the National Centre for the Replacement, Refinement and Reduction of Animals in Research

Preface

There are many pharmaceuticals on the market or undergoing clinical trials that require intravenous infusion, for either short or longer periods, intermittently or continuously, and so this book should be of interest to those in pharmaceutical research and development, as well as in other research areas. These applications include chemotherapy (Skubitz 1997; Vallejos et al. 1997; Ikeda et al. 1998; Patel et al. 1998; Stevenson et al. 1999; Valero et al. 1999) and the treatment of various diseases such as HIV (Levy et al. 1999), hepatitis C (Schenker et al. 1997), and cardiovascular disease (Phillips et al. 1997), as well as during and following problematical surgical procedures (Bacher et al. 1998; Llamas et al. 1998; Menart et al. 1998). It is a regulatory and ethical requirement that these pharmaceuticals first be tested on both rodents and non-rodents by the clinical route, and so the range of pre-clinical experimental models is covered. The technique of prolonged intravenous delivery in conscious, free-moving animal models has also broadened the opportunities to study and evaluate the safety and efficacy of those products that have limiting biological or chemical properties such as half-life and formulation issues.

In 2000 the first book covering pre-clinical infusion techniques was published by Taylor & Francis (*Handbook of Pre-Clinical Continuous Intravenous Infusion*, Healing and Smith, editors) and this has become the singular reference source for this technology up to the present time. However, it is now recognised that a number of the techniques have been refined since that time, and also that new and improved equipment has been developed. In addition, the challenges of providing more complex formulations that are compatible with the infusion models have increased. We therefore decided to produce a more current techniques manual, and have also approached the topic from a fresh angle and structured the chapters differently to provide a novel approach rather than attempting to simply update the original reference book. To provide a complete reference material on this technology, there is now sufficient information to warrant two volumes. Volume I (*Non-Clinical Vascular Infusion Technology: The Science*) covers the scientific principles behind the delivery systems, from both physical and physiological standpoints, and also details formulation-specific considerations. Volume II (*Non-Clinical Vascular Infusion Technology: The Techniques*) covers the practical aspects and methodologies of conducting the studies.

Volume I has been conceived as a consequence of the need to bring together the scientific understanding of all those complex processes that are involved in what first appears to be a 'simple' administration of a xenobiotic directly into central circulation. The forces involved include physico-chemical qualities of both the formulation and blood itself, such as osmolality, pH, viscosity, and surface tension, as well as the various physiological and artificial positive and negative forces responsible for the flow of delivery both internally and externally and the internal distribution of the xenobiotic. An understanding of all these individual elements has been developed and described over many years without previously being brought together in one publication related to the specific vascular delivery systems used both clinically and non-clinically today. This volume will also increase the understanding of the potential effects of formulation vehicles and excipients themselves as well as the dynamics within the body during vascular infusion delivery. Throughout the last decade, considerable progress has been made not only in the development of the animal models of vascular infusion in the procedural and equipment sectors, but also in the understanding of the dynamics and physiology behind this mode of delivery. The physiology of the delivery also involves knowledge of the composition and effects of a wide range of infusates that comprise the modern-day control/vehicle substances commonly used in non-clinical vascular infusion studies. Within pharmaceutical development, the need for more precise and targeted drug delivery has increased in recent years. The regulatory requirement to mimic proposed clinical dosing regimes and modes of delivery in the non-clinical safety program has resulted in the need for some technically demanding practical considerations in order to achieve the regimes without having a deleterious effect on the welfare of the animal involved. It has, therefore, become increasingly important to maintain the reproducibility of the animal models in order to improve the predictability of the effects of infusates when administered to humans. An integral part of the assessment of such products is to have a clear understanding of the potential changes produced by control or excipient substances so that these can be distinguished from those initiated by active ingredients.

The underlying science behind these substances and the control of their effects by understanding the dynamics and physiology of the delivery system is something that has not been adequately addressed in either the *Handbook of Pre-Clinical Continuous Intravenous Infusion* referenced previously or in any single publication. It is, therefore, the purpose of Volume I to provide an understanding of the physiological control of body fluid homeostasis as this relates to vascular infusion technology. This is reflected in sections covering fluid intake and losses, acid/base balance and diffusion, and how these elements may affect the rate and volume of the delivery of the xenobiotic itself. To fully understand the processes involved in an infusion setting, consideration must be given to those substances that commonly form a part of vascular infusion formulations and to their physiological effects in the major laboratory species, and how these changes may be controlled. In laboratory animal models of vascular infusion, the consequences of the delivery are always a function of an over-exposure to a particular element or series of elements of the formulation, such as hypernatraemia, hyperchloraemia, hyperkalaemia, hypercalcaemia, and hyperphosphataemia. These conditions are discussed along with other aspects of formulation technology associated with understanding of successful vascular infusions.

In Volume II, we have taken the opportunity to reflect current expertise by including authors who are the current leaders in the field of commercial infusion technology and, therefore, have the most practical experience (hence also providing the most robust background data sets). It should be noted that many authors' laboratories conduct procedures in more species than those covered in their respective chapter(s), but to gain the widest possible selection of opinions, techniques, and background data, it was necessary to limit each to a specific aspect. There are also variations in the techniques used in different countries, and this has been reflected in the truly international selection of authors. The book is organised by species (those commonly used in pre-clinical studies), namely rat, mouse, dog, minipig, large primate, and marmoset, and there are also chapters covering juvenile studies and reproductive toxicity studies. Each section is organised in a consistent manner to help find the relevant information quickly, and covers information on the selection of the best model, best practice both surgically and non-surgically, practical techniques, equipment selection, and commonly encountered background pathologies.

A specific driver for this book was to identify and share best practices across the industry. It is intended that this book will prevent unnecessary method-development work and hence potentially decrease animal usage, and also provide guidance on choices for the most acceptable methodologies from an animal welfare perspective, which is particularly pertinent when using higher sentient animals such as dogs and primates.

We therefore hope that all non-clinical 'infusionists' find these books to be more than just useful in their pursuance of understanding outcomes in their vascular infusion studies and that they can form the cornerstone of knowledge about this demanding technology and the relevance to safe clinical use of substances via this route of delivery.

References

Bacher H, Mischinger HJ, Supancic A, Leitner G and Porta S. 1998. Dopamine infusion following liver surgery prevents hypomagnesemia. *Magnesium Bulletin* 20: 49–50.

Healing G and Smith D, eds. 2000. *Handbook of Pre-Clinical Continuous Intravenous Infusion*. Boca Raton, Florida: Taylor and Francis.

Ikeda K, Terashima M, Kawamura H, Takiyama I et al. 1998. Pharmacokinetics of cisplatin in combined cisplatin and 5-fluorouracil therapy: A comparative study of three different schedules of cisplatin administration. *Japanese Journal of Clinical Oncology* 28: 168–175.

Levy Y, Capitant C, Houhou S, Carriere I et al. 1999. Comparison of subcutaneous and intravenous interleukin-2 in asymptomatic HIV-1 infection: A randomised controlled trial. *Lancet* 353: 1923–1929.

Llamas P, Cabrera R, Gomezarnau J and Fernadez MN. 1998. Hemostasis and blood requirements in orthotopic liver transplantation with and without high-dose aprotonin. *Haematologica* 83: 338–346.

Menart C, Petit PY, Attali O, Massignon D, Dechavanne M and Negrier C. 1998. Efficacy and safety of continuous infusion of Mononine® during five surgical procedures in three hemophilic patients. *American Journal of Hematology* 58: 110–116.

Patel SR, Vadhanraj S, Burgess MA, Plager C et al. 1998. Results of two consecutive trials of dose-intensive chemotherapy with doxorubicin and ifosfamide in patients with sarcomas. *American Journal of Clinical Oncology* 21: 317–321.

Phillips BG, Gandhi AJ, Sanoski CA, Just VL and Bauman JL. 1997. Comparison of intravenous diltiazem and verapamil for the acute treatment of atrial fibrillation and atrial flutter. *Pharmacotherapy* 17: 1238–1245.

Schenker S, Cutler D, Finch J, Tamburro CH et al. 1997. Activity and tolerance of a continuous subcutaneous infusion of interferon-alpha-2b in patients with chronic hepatitis C. *Journal of Interferon and Cytokine Research* 17: 665–670.

Skubitz KM. 1997. A phase I study of ambulatory continuous infusion of paclitaxel. *Anti-Cancer Drugs* 8: 823–828.

Stevenson JP, DeMaria D, Sludden J, Kaye SB et al. 1999. Phase I pharmacokinetic study of the topoismerase I inhibitor GG211 administered as a 21-day continuous infusion. *Annals of Oncology* 10: 339–344.

Valero V, Buzdar AU, Theriault RL, Azarnia N et al. 1999. Phase II trial of liposome-encapsulated doxorubicin, cyclophosphamide, and fluorouracil as first-line therapy in patients with metastatic breast cancer. *Journal of Clinical Oncology* 17: 1425–1434.

Vallejos C, Solidoro A, Gomez H, Castellano C et al. 1997. Ifosfamide plus cisplatin as primary chemotherapy of advanced ovarian cancer. *Gynecologic Oncology* 67: 168–171.

Acknowledgements

In a publication such as this, along with the primary contributors there are usually a number of individuals supporting it in many ways. This publication is no exception.

Both principal authors, Dr Owen P Green and Dr Guy Healing, owe a debt of gratitude to our employers, LeVerts Ltd and AstraZeneca, respectively, for the time that we were given for the project as well as the administrative support.

We would also like to thank our co-authors in this particular volume for their scientific input and technical knowledge that has been invaluable in creating this publication:

James RA Baker (AstraZeneca, UK)
Dean Hatt (GlaxoSmithKline, UK)
Sophie Hill (AstraZeneca, UK)
Kevin Sooben (AstraZeneca, USA)

Such a labour over a long period to produce a book of this standard also requires the support of those who also end up sacrificing some of their personal time whilst the authors slave over their manuscripts. To this end we thank our long-suffering families for their patience and fortitude.

We would also like to express our appreciation to the laboratories that have provided some essential background data for various species that have been used in this volume:

AstraZeneca
Charles River Laboratories
Covance Laboratories
Huntingdon Life Sciences
WIL Research

chapter one

Body fluid dynamics

The objective of this chapter is to bring together the
principles that are known about fluid requirement
and management by the mammalian body under
normal conditions and how this might be affected by
typical vascular infusion conditions in non-clinical
toxicology studies. Since the introduction of fluid
or indeed the withdrawal of fluid from centralised
circulation is inevitably going to disturb the physiological fluid balance in a normovolaemic subject, it
would be important under these circumstances to
understand the principles of body fluid homeostasis. In this chapter an overview of the principles of
body fluid intake, storage, movement, distribution,
and elimination will be given in order to provide a
means of assessing acceptability of infusion conditions. The principal relevance to vascular infusion
studies in laboratory animals is one of fluid overload. The administration of a xenobiotic undergoing
safety assessment via vascular infusion delivery is
normally undertaken in normovolaemic animal
models; consequently it is important to have knowledge of what effects and changes may result from
increasing the fluid load during such investigations.

Introduction

In the late nineteenth century, Ernest Starling proposed the concept
that fluid exchange across blood vessels was governed by the balance
between hydrostatic and osmotic pressure gradients between the intravascular and interstitial fluid compartments (Starling 1896). A hydrostatic pressure gradient in excess of the osmotic gradient at the arterial
end of the capillary bed results in a net transudation of fluid into the
interstitium. At the venous end of the capillary bed, plasma proteins,
which cannot pass out of the blood vessels, exert an osmotic force in
excess of the hydrostatic gradient, resulting in a net fluid flux into

the vessels. Well over a century of research has confirmed that Starling's hypothesis provides the foundation for microvascular fluid exchange. However, it has revealed that the necessary physiology and anatomy are much more complex. Consequently, a much deeper understanding of transvascular fluid dynamics is necessary for a logical and rational approach to intravenous infusion technology involving fluids containing xenobiotics.

Composition and units of measurement

Before considering the potential effects of excess volume and concentration of solutes in intravenous infusion delivery settings, it is important to review the basic constituents of body fluids in mainstream laboratory animal species (Table 1.1).

The molecular weight of a compound is defined as the sum of all the atomic weights of elements specified in the chemical formula of the compound. One mole (mol) of any substance is the molecular weight of the substance in grams, and one millimole (mmol) is 10^{-3} mol, or the molecular weight of a substance expressed in milligrams (mg). The concentrations of these charged solutes would be relatively small and are usually expressed in millimoles per litre (mmol/L).

Table 1.1 Molecular Weights of Some Physiological Constituents of Body Fluids

Substance	Symbol/formula and charge/valence	Molecular weight
Sodium	Na^+	23
Potassium	K^+	39.1
Calcium	Ca^{2+}	40.1
Magnesium	Mg^{2+}	24.3
Oxygen	O	16
Nitrogen	N	14
Hydrogen	H^+	1
Chloride	Cl^-	35.5
Phosphate	PO_4^{3-}	95
Bicarbonate	HCO_3^-	61
Sulphate	SO_4^-	96.1
Carbon dioxide	CO_2	44
Lactate	$C_3H_5O_3^-$	89
Glucose	$C_6H_{12}O_6$	180
Urea	NH_2CONH_2	60
Water	H_2O	18

Concentrations of ions in biological solutions are often quoted in terms of electrochemical equivalence, milliequivalents per litre (mEq/L). One equivalent is defined as the weight in grams of an element that combines with or replaces 1 gram of hydrogen ion (H^+). Since 1 gram of H^+ is equal to 1 mole of H^+, 1 mole of any univalent anion (charge equals 1^-) will combine with this H^+ and is equal to 1 equivalent (Eq).

These measurements of concentration of body fluid constituents can be related by the following equations:

$$mEq/L = mmol/L \times valence$$

$$mEq/L = \frac{mg/dL \times 10}{MW} \times valence$$

According to Avogadro's law, regardless of its weight, one mole of any substance contains the same number of particles (6.023×10^{23}). Solutes exert an osmotic effect in solution that is dependent only on the number of particles in solution (see Chapter 2).

Compartmentalisation and distribution

Body fluids are distributed intracellularly and extracellularly. Extracellular fluid (ECF) comprises blood plasma and interstitial fluid, the latter being outside the vascular compartment. Cerebrospinal fluid, gastrointestinal fluid, lymph, bile, glandular and respiratory secretions, and synovial fluid are in equilibrium with other extracellular fluids but are contained in specialised compartments. These fluids are not transudates of plasma, but instead are produced by the action of specific cells and consequently are called *transcellular fluids* (Edelman et al. 1952). Water, Na^+, K^+, and Cl^- in the transcellular fluids are readily exchangeable with ECF (Edelman and Sweet 1956; Nadell et al. 1956). As the body fluid spaces are traditionally conceptualised anatomically, the capillary endothelium and the cell membranes are the boundaries of the vascular space and the ECF compartment, respectively. Functionally, it would be more correct to view these body fluid spaces as volumes of water and electrolytes in dynamic equilibrium, since fluids and ions move across these semi-permeable boundary membranes.

The approximate sizes of the total body water (TBW), ECF, and intracellular fluid (ICF) spaces reflect the partitioning of body fluids and solutes. Loss or gain of fluids or solutes from one of these compartments may result in alterations of the volumes of other compartments. The ability to predict relative changes in fluid compartment size is important in understanding the pathophysiology of the vascular delivery of substances in healthy subjects and consequently the interpretation of any untoward effects of those substances. Fluids administered parenterally initially enter the ECF, and it is, therefore, important to estimate the size of the ECF space for the animal model being used.

The volume of a body fluid compartment has often been assessed by using a known concentration and volume of an indicator solution when:

$$V_d = \frac{\text{total dose of indicator}}{\text{concentration of indicator after mixing}}$$

where V_d represents the volume of distribution.

Various indicators have been used to estimate ECF, TBW, plasma volume (PV), and erythrocyte volume (see Table 1.2), although these figures can be very variable across and within species. The ICF volume would then be the difference between the TBW and the ECF.

Total body water volume is often defined as 60% of body weight, but there is considerable individual variation in this value. In humans, TBW declines with age and is lower in women than in men (Edelman and Leibman 1959). This is most likely explained by differences in body fat content since fat is lower in water content than lean tissue. Therefore, it is considered that TBW constitutes about 60% of body weight in adult mammals that are not obese. It is important when estimating the TBW compartment and the potential effects of infusion volumes and rates of delivery that this is made on the basis of lean body mass. The following formulas are proposed for estimating lean body mass on the basis of the assumptions that (a) approximately 20% of body weight is due to fat in normal animals, (b) morbid obesity represents an increase in body fat to at least 30% of body weight, and (c) body weight is a reasonable estimate of lean body mass in thin animals:

Obese   Body weight × 0.7 = Lean body mass
Normal   Body weight × 0.8 = Lean body mass
Thin   Body weight × 1.0 = Lean body mass

An estimate of ECF volume expressed as a percentage of body weight would also be useful for clinical assessment of fluid shifts resulting from infusion delivery regimes as a basis for determining potential adverse effects from the delivery regime. For this parameter, as for the others, the data derived are very variable as a consequence of the various methods used to estimate this value and because the space itself is heterogeneous (Edelman and Leibman 1959). Four sub-compartments of the ECF have been defined: plasma, interstitial lymph,

Table 1.2 Estimates of Erythrocyte, Plasma, Whole Blood, and Extracellular Fluid Volumes

Species	Erythrocyte volume (mL/kg)	Plasma volume (mL/kg)	Whole blood volume (mL/kg)	Total body water (mL/kg)	ECF (mL/kg)	References
Human (60 kg)	23.2–26.8	37.0–39.8	67.0–71.0	446–646	115–209	Taggart et al. 1989; Grollman et al. 1953; Retzlaff et al. 1969; Cameron et al. 1999; Watson et al. 1980
Rat (250 g)	19.5–24.4	36.0–52.7	57.9–73.0	669–693	202–228	Check et al. 1957; Foy and Schnieden 1960; Lee and Blaufox 1985
Mouse (25 g)	28.2–36.5	55.0–68.3	83.5–99.0	585–800	188–222	Sheng and Huggins 1979; Riches et al. 1973
Dog (10 kg)	31.4–44.6	42.0–65.0	83.0–114.0	598–654	201–246	Baker and Remington 1960; Woodward et al. 1968; Altman and Dittmer 1974; Walser et al. 1953; Becker and Joseph 1955; Zweens J et al. 1978
Minipig (10 kg)	23.0–34.0	33.6–37.4	61.0–68.0	497–617	680–1312	DeBruin 1997; Hawk et al. 2005; Jorgensen et al. 1998
NHP (2.5 kg)	25.5–29.0	42.5–60.0	75.0–85.0	623–657	81.2–97.0	Azar and Shaw 1975; Kamis and Norhayatee 1981; Zia-Amirhusseini et al. 1999

dense connective tissue, and transcellular compartments. The relative proportions of these compartments when compared to body weight are shown in Figure 1.1.

Water lost during fluid withdrawal or fluid gained during infusion delivery may equilibrate more slowly with some sub-compartments of the ECF, such as the dense tissue, in comparison with plasma itself. Studies in dogs with radiolabelled materials have demonstrated equilibrium, including bone water, to be quite rapid at 2 to 4 hours (Edelman et al. 1952; Edelman et al. 1954), with the rate-limiting step being the transfer across

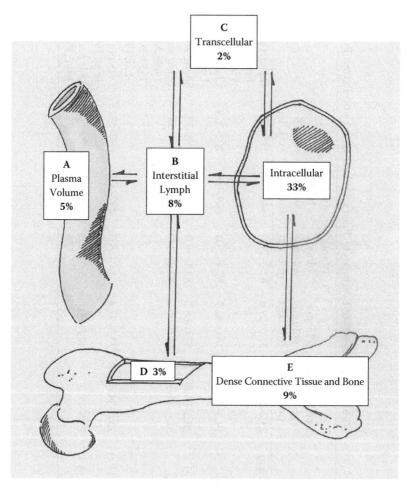

Figure 1.1 (See colour insert.) Compartments of total body water expressed as percentage of body weight. A+B+C+D+E = 27% Body Weight (Total extracellular fluid volume); A+B+C+D = 18% Body Weight (Rapidly equilibrating space).

cell membranes. From a physiological perspective and for the purpose of determining fluid volume tolerances in animal infusion studies, ECF should be estimated as about 27% of lean body weight.

Before understanding the distribution of solutes around the fluid compartments, it is important to have data on the normal concentration of solute components within the 'resting' compartments and in particular the plasma or ECF compartment (Table 1.3).

Movement between compartments/exchange

Movement of fluid between intracellular and extracellular compartments is determined by the number of osmotically active particles in each space. Sodium (Na^+) and its associated anions account for most of the osmotically active particles in the ECF together with significant contributions from glucose and urea. The ECF osmolality may be estimated from the following formula (Rose 1984):

$$\text{ECF osmolality (mOsm/Kg)} = 2(Na^+ + K^+) + \frac{\text{Glucose}}{18} + \frac{\text{BUN}}{2.8}$$

Cell membranes are permeable to urea and K^+. Therefore, these solutes are ineffective osmoles. Effective osmolality is calculated as

$$\text{Effective ECF osmolality (mOsm/Kg)} = 2 \times Na^+ + \frac{\text{Glucose}}{18}$$

In most laboratory animal species, the normal blood glucose concentrations are small (70–110 mg/dL) and do not contribute significantly to effective ECF osmolality. Therefore, $2 \times [Na^+]$ is a good approximation of the effective ECF osmolality.

All body fluid spaces are isotonic with one another. Thus the effective osmolality of the ICF may also be estimated by doubling the ECF Na^+ concentration even though this is small in the ICF. Because of isotonicity of all body fluid spaces, it is postulated that the tonicity of the TBW may also be approximated by doubling the plasma $[Na^+]$. The relationship, unsurprisingly, is that when total exchangeable Na^+ increases, the plasma sodium concentration also increases (Rose 1989), and these changes are usually associated with body fluid hypertonicity, a typical response of intravenous infusion of 0.9% NaCl-based formulations. As can be predicted, an infusion of NaCl will result in an increase in the displacement of K^+ from cells in order to avoid a hyponatraemia.

Table 1.3 Plasma Concentrations of Electrolytes[a]

Species[b]	Na⁺ (mEq/L)	K⁺ (mEq/L)	Ca²⁺ (mEq/L)	Mg²⁺ (mEq/L)	Cl⁻ (mEq/L)	HCO₃⁻ (mEq/L)	HPO₄²⁻ (mEq/L)	Proteins (g/dL)
				Electrolyte concentration				
Human[b]	135–145	3.5–5.0	2.3–3.6	1.6–2.1	96–106	23–30	1.4–2.2	6.0–8.5
Rat	137–149	2.5–4.8	5.0–6.1	1.3–2.1	97–110	26–28[c]	2.4–6.2	5.9–8.9
Mouse	147–151	4.5–6.4	2.3–2.8	1.5–2.0	107–111	19–24[d]	1.4–4.8	4.9–5.8
Dog	144–151	3.7–4.9	4.9–5.7	1.3–1.6	105–115	17–24	2.0–3.5	5.0–6.3
Minipig	140–149	4.9–7.9	2.6–3.0	nda	96–104	27–31[e]	3.3–4.8	6.3–7.8
NHP	145–154	3.4–5.3	2.3–2.7	nda	102–110	15–23[f]	4.2–6.1[g]	7.1–9.0

[a] Background data supplied by various contract research laboratories (5–95 percentiles).
[b] Encyclopedia of Nursing and Allied Health.
[c] Kent Webb et al. 1977.
[d] Kovacikova et al. 2006.
[e] Guzel et al. 2012.
[f] Hobbs et al. 2010.
[g] Sandeers-Beer et al. 2011.
nda: no data available.

Body fluid homeostasis

Exchange of fluid between plasma and interstitial spaces

The partitioning of fluid between plasma and interstitial fluid spaces is critically important for the maintenance of the effective circulating blood volume and overall body fluid homeostasis during deviations from 'normal' conditions, particularly the excess volumetric state incurred as a consequence of intravenous infusion regimes in non-clinical studies. The effective blood volume can be described as 'the component of blood volume to which the volume-regulatory system responds by causing sodium and water retention in the setting of cardiac and hepatic failure even though measured total blood and plasma volume may be increased' (Peters 1948; Schrier 1988). Approximately 18% of the ECF (see Figure 1.1) is contained in the plasma volume. Exchange of sodium and fluid between plasma and interstitial spaces occurs at the capillary level. The volume of the vascular space is controlled by a balance between forces that favour filtration of fluid through the vascular endothelium (capillary hydrostatic pressure and tissue oncotic pressure) and forces that tend to retain fluid within the vascular space (plasma oncotic pressure and tissue hydrostatic pressure). Oncotic pressure is the osmotic pressure generated by plasma proteins in the vascular space. These relationships are described by Starling's law (see Figure 1.2).

$$\text{Net filtration} = K_f \left[(HP_{cap} - HP_{if}) - (OP_p - OP_{if}) \right]$$

where K_f represents the net permeability of the capillary wall; HP represents the hydrostatic pressure generated by the heart (HP_{cap}) or interstitial fluid (HP_{if}); OP represents the oncotic pressure (colloid osmotic pressure) generated by the plasma proteins (OP_p) or filtered proteins and mucopolysaccharides in the interstitium (OP_{if}).

At the venule end of the capillary, the forces favouring filtration are less than the forces favouring reabsorption of fluid into the vascular space because capillary hydrostatic pressure decreases along the length of the capillary but capillary oncotic pressure remains the same (Rose 1984). Some of the fluid that is filtered into the interstitium at the proximal end of the capillary is reabsorbed distally, and the remainder of the filtered fluid is transported via the lymph system in the interstitium. The hydrostatic pressure transferred from arterioles to the capillaries is controlled by autoregulation of the precapillary sphincter, which protects the capillary from increases in hydrostatic pressure caused by systemic hypertension as a consequence of increases in intravenous fluid volume during infusions. Consequently, up to a point, dangerous loss of vascular fluid into the interstitial fluid by filtration during excess intravenous infusion delivery can be compensated.

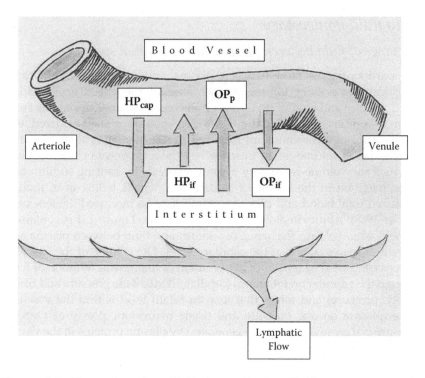

Figure 1.2 (See colour insert.) Factors affecting fluid movement at the capillary level. HP_{cap} = capillary hydrostatic pressure; HP_{if} = interstitial hydrostatic pressure; OP_p = capillary oncotic pressure; OP_{if} = interstitial oncotic pressure.

During periods of fluid excess, capillary oncotic pressure decreases and hydrostatic pressure may increase if the excess is severe enough to cause hypervolaemia. This could often be the case during vascular infusion studies with isotonic formulations. These alterations in Starling's forces favour an increase in net filtration of fluid into the interstitium at the level of the capillary. Decreased reabsorption of interstitial fluid augments effective circulating blood volume, thus increasing plasma protein concentration and decreasing hydrostatic pressure. Conversely, increases in plasma protein will increase plasma oncotic pressure and reduce the net force favouring filtration of fluid out of the capillary. In the healthy, normovolaemic animal, maintenance of plasma volume depends on a fine balance between the forces favouring filtration and those favouring reabsorption in the capillary. Therefore, as long as there are no consequential effects on renal function during a toxicology study by the vascular infusion route, renal clearance of excess fluid volume will maintain homeostasis, although some minor increases in interstitial fluid may be experienced with associated intermittent lowering of plasma electrolyte concentrations.

Zero balance

In the healthy adult animal at rest in a controlled environment, daily intake of water, nutrients, and minerals is exactly balanced by daily excretion of these substances or their metabolic by-products. Thus, the animal does not experience a net gain or loss of body water, nutrients, or minerals and is said to be in zero balance, fluid intake being equal to fluid loss (see Figure 1.3). The volume of fluid added to body fluids via food and water consumption, and by metabolism, is equal to the volume of fluid

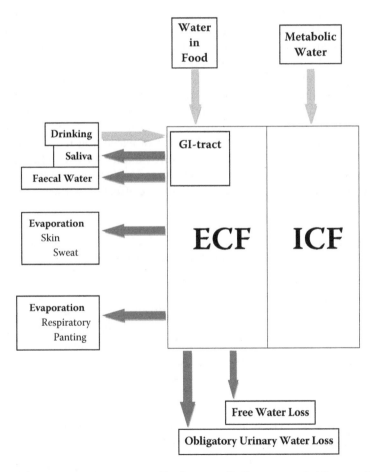

Figure 1.3 (See colour insert.) Total body water. Daily input and obligatory losses.

lost by various processes such as via urine, faeces, saliva, and evaporation from cutaneous and respiratory epithelia (Maxwell et al. 1987).

To get some measure of zero balance in laboratory animals, it is routine to look comparatively at water consumption and urinary output. It is difficult to measure water losses in saliva and faeces, which are small under normal conditions. Although evaporative losses may be great in heat-stressed, exercising, or active animals, the most important obligatory daily loss of water in healthy animals in a controlled environment is via the urine. Values for water intake by drinking and water loss via urine in various laboratory animal species is given in Table 1.4.

A volume of fluid considered essential to maintain an animal in zero fluid balance could be termed the *maintenance fluid need* and would be dependent on many factors such as water losses, ambient temperature and humidity, voluntary or forced activity of the animal, and the animal's physical condition. In toxicology studies we can assume that the environmental conditions under which the animals are maintained are fairly uniform and are well controlled, whereas the physical condition of the animal will be dependent on the outcome of the study itself. This is largely an unknown going into such a study. However, with vascular infusion studies, the volume and rate of delivery of the infusate is predetermined and, therefore, can be adjusted to complement the physiological circumstances. Therefore, vascular infusion into normal healthy animals, by definition, creates a situation of exceeding the maintenance fluid need and disrupts the concept of zero balance for which the animals will need to compensate. The concomitant water intake requirements during a vascular infusion study will also be affected by the composition

Table 1.4 Estimated Measurements of Daily Water Intake and Output in Humans and Laboratory Animals

Species	Input – Water intake (mL/kg/day)	Output – Urine volume (mL/kg/day)	References
Human (60 kg)	38.5–78.0	16.7–33.3	Putnam 1971; US Daily Reference Intake Values
Rat (250 g)	138.6–169.4	13.2–117.2	US EPA 1988
Mouse (25 g)	102.9–279.4	42.8	Bernstein 1968; US EPA 1988
Dog (10 kg)	19.5–84.0	10.5–17.9	O'Connor and Potts 1969
Minipig (10 kg)	60.0–80.0	24.0	Barnett 2007
NHP (2.5 kg)	17.0–97.0	30.0–212.0	*Nutrient Requirements of Nonhuman Primates* 2003; Suzuki et al. 1989

of the infusate. The amount of solute in the infusate will affect the volume of urine required for solute excretion, water intake, and urinary osmolality. Infusates with higher solute contents require a greater total water intake than infusates of relatively lower solute content. Most animals in pre-clinical studies have free access to water and, therefore, ingest sufficient water to support urinary excretion of excess solute provided by vascular infusion.

Fluid losses

The principal mechanism of voiding excess fluid from the body is urine production. Other mechanisms such as faeces production and respiratory and cutaneous evaporative losses are minor in comparison and not considered significant in an acute fluid excess condition induced by vascular infusion delivery. Consequently, this section will focus on the role of urine production in the voidance of excess fluid under these conditions.

Daily urinary water losses can be classified as obligatory water loss (water needed to excrete the daily renal solute load) and free water loss (water excreted unaccompanied by solute under the control of antidiuretic hormone [ADH]). Clearance of free water increases during relative fluid excess, thus protecting the animal from the over-hydration and hypotonicity that would result from retention of water in excess of solutes. Obligatory renal water loss must occur even in states of relative water deficit so that solute may be eliminated from the body.

Solute control – Obligatory urinary water losses

The amount of fluid required for elimination of urinary solute load in theory depends on the maximum urine osmolality that can be achieved by the animal (Table 1.5). However, to complicate matters, solute is usually not excreted at maximum urine osmolality, especially when water is readily available for voluntary consumption or when normal hydration

Table 1.5 Estimated Mean Urinary Osmolalities

Species	Osmolality (mOsm/kg)	References
Human (60 kg)	1200 ± 250	Eastwood 1997
Rat (250 g)	1450 ± 350	Sands et al. 1985; Salas et al. 2004
Mouse (25 g)	1490 ± 360	Bernstein 1968
Dog (10 kg)	1541 ± 527	van Vonderen et al. 1997
Minipig (10 kg)	843 ± 44	Shaw et al. 2006
NHP (2.5 kg)	632 ± 449	*Contemporary Topics*, Vol. 42

levels are exceeded during infusion deliveries. In experiments in dogs (Hardy and Osborne 1979; O'Connor and Potts 1988; O'Connor and Potts 1969; Chew 1965), increased renal solute induced by increased oral fluid intake resulted in increased urine volume; however, urinary osmolality remained relatively normal at approximately 1600 mOsm/ kg. Urine osmolalities did not, as might be expected, increase toward the maximum attainable in water-deprived dogs. Thus urinary osmolality is conserved in the presence of increased urine solute load by an increase in urine volume. As well as with increased fluid intake via the oral route, this is the expected consequence of any vascular delivery of excess volume and can be detected with an appropriately timed urine collection.

The renal solute load is usually determined by normal dietary and water intake levels of such elements as urea, Na^+, K^+, Ca^{2+}, Mg^{2+}, NH_4^+ plus other cations, and PO_4^{3-}, Cl^-, SO_4^{2-} plus other anions; but, in the case of vascular infusion of fluids, we have the additional quantities of such elements in the infusate formulation. Since in normal, healthy laboratory animals the dietary intake and water intake are well controlled, it is expected that the majority of solute given by way of a vascular infusion would be excreted in the urine.

Volume control – Urinary free water

Excretion of urinary free water is controlled by suppression of the secretion of antidiuretic hormone (ADH). In the case of vascular infusion, much is dependent on the solute composition of the infusate, but, in the main, this mode of delivery will result in an inhibition of ADH. Most solutes will be designed to be as isotonic as possible, thereby reducing the likelihood of increases in urinary free water. However, it is not uncommon for water to be ingested by animals in intravenous infusion studies at only slightly reduced rates when compared to normal conditions. This would result in the dilution of body solute and hypotonicity. Only small changes in serum osmolality are needed to inhibit secretion of ADH (Robertson 1983; Robertson et al. 1976). Consequently, daily urinary free water losses would be large during fluid intake excess in otherwise healthy animals.

Fluid intake

The natural intake of fluid (water) by animals on a daily basis is considered to be related to food intake in normal healthy individuals and has been demonstrated in dogs and rats (Cizek 1959; Chew 1965; Cizek and Nocenti 1965). In various studies, limiting the food intake of animals will have the effect of reducing their natural water intake; however, a confounding factor would be the potential moisture content of the diet fed to the animals. This is one reason why dietary provisions for laboratory animals on study are very well controlled. It is well known that the percentage of water in animal feeds is variable (Lewis and Morris 1987). As a result, not only are standard diets used with well-controlled content but also the amounts fed at regular intervals follow discrete patterns related to the growth schedule of the species.

It is not the intention of this text to discuss in detail the dietary needs of laboratory animals and the effect this might have on fluid loading and physiological changes in intravenous infusion dosing toxicology studies. Rather, it is accepted that dietary intake for laboratory animals is well controlled with the use of composite dry foods, which has the effect of normalising water intake in healthy control animals. However, within the context of this discussion, the normal drinking water intake of laboratory animals (Table 1.6) is worthy of record.

The data in Table 1.6 is extrapolated from what we know of the established average daily water intake and urinary output of the major laboratory animal species. The data can be expressed as a ratio of fluid intake to obligatory fluid loss, which reveals some interesting comparisons between the species and potential projections on the tolerability of the various species to fluid loading during vascular infusion scenarios. Whilst the smaller mammals such as rat and mouse have much higher fluid requirements per body weight than humans, the resultant obligatory fluid loss is within similar proportion to humans. Conversely, the non-human primate appears to have an extraordinary high obligatory fluid loss in comparison with its daily water intake. It is likely that this apparent difference is inherently linked to the difficulty in accurately determining the actual water intake in this species but also has to do with the high water content of the dietary supplements given. In any event, it is most likely that the non-human primate is the laboratory animal species at most variance from the human in this regard. When it comes to assessing the effects that this physiological capability would have on venous infusion of fluids, it could be concluded that, whilst most species are similar to

Table 1.6 Estimated Normal Daily Water Intakes and Fluid Loss

Species	Water intake (mL/kg/day)	Obligatory fluid loss (mL/kg/day)	Ratio
Human	58	25	2.32
Rat	154	65	2.36
Mouse	191	43	4.44
Dog	52	14	3.71
Minipig	70	24	2.92
NHP	57	110	0.52

humans when assessing the ratio of obligatory fluid loss to water intake, it is possible that the species with the higher obligatory fluid losses would be more tolerant to circulating physiological fluid excesses as experienced during infusion studies in normovolaemic animals.

The variation in daily water intake for all species is also influenced by the solute load of the diet. Approximately 66% of the renal solute load is urea, an end product of protein metabolism, and increasing the protein content of the diet increases the renal solute load. Diets high in protein are also associated with greater total water intake, but it would be rare for an intravenous infusion formulation to have such constituents and thus affect water intake in what would be considered a converse way in relation to the imposed increased fluid load. However, the ions Na^+, K^+, Ca^{2+}, Mg^{2+}, PO_4^{3-}, Cl^-, and SO_4^{2-} also contribute to dietary solute, and increasing percentages of salt in foods is associated with increased water intake (Burger et al. 1980). This would also be true should the increased salt be associated with an intravenous infusion formulation.

Maintenance of body fluid balance

Body fluid balance is complex even in healthy laboratory animals in a thermoneutral environment and consistently receiving standard controlled dietary feeds, and some of the more important features of intravenous fluid delivery that may affect this homeostatic relationship have been discussed. In conclusion, despite the intravenous infusion of large volumes of infusate over and above the normovolaemic condition of the animal model, the animal will continue to drink a modified amount of water daily based upon the animal's body weight and the renal solute load in the infusate (assuming the dietary load is controlled and constant) and, of course, the consequent obligatory water losses for urinary solute excretion. At this stage it would be interesting to be able to calculate the required amount of water for each of the major laboratory animal species under these normal laboratory conditions.

From a veterinary point of view, maintenance fluid needs in dogs have been defined as 60 mL/kg/day in smaller dogs and 40 mL/kg/day in the larger dogs (Muir and DiBartola 1983). However, for purposes of determining fluid needs in conscious, healthy, fully ambulatory animals, some relationship with energy needs, food intake, and body weight would need to be taken into consideration. Not surprisingly, therefore, fluid maintenance needs for dogs have been assessed on the basis of calorific needs (Harrison et al. 1960; Haskins 1984). Normal maintenance energy requirement can be defined as the number of calories required to sustain the basal metabolic rate, which would include energy for digestion, absorption, food conversion, maintenance of body temperature, and normal activity. Maintenance of such energy usage has been expressed in the following formula (Brody et al. 1934):

$$140 \text{ kcal} \times \text{body weight (kg)}^{0.73}$$

Therefore, a 10-kg dog would require 750 kcal of energy per day, or 75 kcal/kg/day. Following another ancient publication (Adolf 1939), where mean water intake was approximated to 1 mL/kcal, the water requirement would be 75 mL/kg/day. The physiological reasons for the correlation between calorific and water needs are not well documented, and there is considerable variation in opinions on the formula for calculating basal energy requirements. The normal daily water intakes as given in Table 1.6 are probably of most use when estimating total fluid requirement in normal laboratory conditions, but comparisons with the modified values for calorific requirement are of interest particularly in the modern ambulatory vascular infusion models. For dogs, it is generally accepted that maintenance fluid needs approximate to a 2-fold factor of the basal fluid requirements, which is in line with the maintenance and basal calorific energy needs (Abrams 1977; Rivers and Burger 1989) and, whilst opinions vary, the most popular calculation can be represented by the following formula:

$$\text{Basal energy requirement} = 97 \times \text{body weight (kg)}^{0.655}$$

So, based on this formula, a 10-kg dog has a basal energy requirement of 44 kcal/kg/day and, hence, a maintenance requirement of 88 kcal/kg/day. As before, assuming 1 mL of water required per kcal of energy need, the maintenance fluid requirement for this dog would be approximately 88 mL/kg/day and the basal requirement 44 mL/kg/day. This is clearly not a precise figure but more an estimation that has merits when applied to laboratory animals. Whilst similar formulae have not been formally established for other laboratory animal species, it is interesting to apply this same formula to the other species for comparison with direct measurements of fluid intake (Table 1.7). Whilst there are many differences in the metabolic needs between rodent and non-rodent species, as well as intra-species variation, for each species in the laboratory great attempts are made from the dietary, environmental, and housing perspectives to normalise this variation. The physiological reason for the correlation between calorific and fluid needs is most likely related to the diet. Fluid intake is in part a function of renal solute load, which, like general calorific requirement, is related to diet, in both quantity and composition.

Several points can be taken from the data presented in Table 1.7, the calculated maintenance water requirements based on calorific needs, when compared with the direct measurement of daily water intakes

Table 1.7 Calculated Daily Water Intakes Based on Calorific Needs

Species	Basal energy requirement (kcal/day)	Basal water requirement (mL/day)	Maintenance energy requirement (kcal/day)	Maintenance water requirement (mL/day)	Calculated maintenance water requirement (mL/kg/day)
Human (60 kg)	1257.8	1258	2515.6	2516	42
Rat (250 g)	38.8	39	77.6	78	312
Mouse (25 g)	9.7	10	19.4	19	633
Dog (10 kg)	438.3	438	876.7	877	88
Minipig (10 kg)	438.3	438	876.7	877	88
NHP (2.5 kg)	176.7	177	353.5	354	142

given in Table 1.6. Within the expected variation in the rather inaccurate means of measuring water intake of the various species, it can be considered that the calculated maintenance water requirement is similar for all the larger species but is significantly higher for the smaller mammals (rat and mouse). The implications for vascular infusion into normovolaemic animals is that, potentially, the rat and mouse can, physiologically, accommodate higher volumes of infusate delivery in comparison with human, dog, minipig, and primate. These are figures to be aware of when considering the capabilities of selected species for vascular infusion studies and that would have an impact on the rates and volumes of delivery discussed in Chapter 3.

Summary

When summarising the effects of fluid control on the vascular infusion delivery of xenobiotics in laboratory animals, the first fact to remember is that the model will be in normovolaemic animals, that is to say, in animals that will have normal variance in measured whole blood volume (mL/kg body weight). This means that all external influences on fluid intake will be controlled. The laboratory environment and, therefore, body temperature will be maintained such that there will be no extreme demands made for water intake due to high temperatures. Most animal facilities are controlled at around 20°C ± 2°C, and this will be sufficient to ensure normal diurnal fluctuations in demand for water intake based on normal activity levels and food intake. The influence of food consumption is also controlled as much as possible, particularly for the small laboratory animals—rats, mice, and rabbits—for which standard feeds are routinely used with consistent levels of constituents, including water content. The effect of food consumption on water intake for the larger animals such as primates and dogs is less controlled as a consequence of the varied and larger water content of their normal dietary constituents. In order to be sure that the animal model used in a toxicology study involving delivery via vascular infusion will tolerate the proposed volume of infusate, it is important to know that the normal fluid balance for the animal will be maintained. Therefore, regular measurements of water intake and urinary volume output are essential during infusion studies to ensure that the basal control of ECF and total blood volume is maintained. These data can then be compared with contemporary control data (similar to data in Table 1.6) so that effects of infusate composition and volume can be assessed and any adjustments made. Ideally, for the long-term delivery of an infusate to be considered 'physiological', the disturbance in body fluid homeostasis should be minimal and can be assessed by a comparison of urinary osmolality values with basal levels (Table 1.5).

References

Abrams JT. 1977. The nutrition of the dog. In: *CRC Handbook Series in Nutrition and Food*, Section G: *Diets, Culture Media and Food Supplements*, Rechcigl M (ed.), p. 1. Boca Raton, Florida: CRC Press.

Adolf EF. 1939. Measurements of water drinking in dogs. *Am. J. Physiol.* 125: 75.

Altman PL and Dittmer DS. 1974. *Biology Data Book*, 2nd edition, pp. 1846–1992. Bethesda, Maryland: Federation of American Societies for Experimental Biology.

Azar E and Shaw ST. 1975. Effective body water half-life and total body water in rhesus and cynomolgus monkeys. *Can. J. Physiol. Pharmacol.* 53(5): 925–939.

Baker CH and Remington JW. 1960. Role of the spleen in determining total body haematocrit. *Am. J. Physiol.*198: 906.

Barnett SW, ed. 2007. *Manual of Animal Technology*. Oxford: Blackwell Publishing.

Becker EL and Joseph BJ. 1955. Measurement of extracellular fluid volumes in normal dogs. *Am. J. Physiol.* 183: 314.

Bernstein SE. 1968. Physiological characteristics. In: *Biology of the Laboratory Mouse*, 2nd edition. Green EL (ed.), New York: Dover Publications Inc.

Brody S, Procter RC and Ashworth US. 1934. *Growth and Development with Special Reference to Domestic Animals XXXIV: Basal Metabolism, Endogenous Nitrogen, Creatinine and Neutral Sulphur Excretions as Functions of Bodyweights.* Columbia: University of Missouri.

Burger IH, Anderson RS and Holme DW. 1980. Nutritional factors affecting water balance in the dog and cat. In: *Nutrition of the Dog and Cat*, Anderson RS (ed.), p. 145. Oxford: Pergamon Press.

Cameron JR, James GS and Roderick MG. 1999. In: *The Physics of the Body*, 2nd edition, p. 182. Madison, Wisconsin: Medical Physics Publishing.

Cheek DB, West CD and Golden CC. 1957. The distribution of sodium and chloride and the extracellular fluid volume in the rat. *J. Clin. Invest.* 36(2): 340–351.

Chew RM. 1965. Water metabolism of mammals. In: *Physiologic Mammology*, Vol. II: *Mammalian Reaction to Stressful Environments*. Mayer WW and Van Gelder RG (eds), p. 43. New York: Academic Press.

Cizek LJ. 1959. Long term observations on relationship between food and water ingestion in the dog. *Am. J. Physiol.* 197: 342.

Cizek LJ and Nocenti MR. 1965. Relationship between water and food ingestion in the rat. *Am. J. Physiol.* 208: 615–620.

DeBruin NC. 1997. Body composition and energy utilisation during the first year of life. Doctoral thesis, Erasmus University. ISBN 9090106774.

Eastwood M. 1987. Water, electrolytes, minerals and trace elements, Chapter 8. In: *Principles of Human Nutrition*. London: Chapman and Hall.

Edelman IS, Olney JM, James AH et al. 1952. Body composition: Studies in the human being by the dilution principle. *Science* 115: 447.

Edelman IS, James AH, Baden H et al. 1954. Electrolyte composition of bone and the penetration of radiosodium and denterium oxide into dog and human bone. *J. Clin. Invest.* 33: 122.

Edelman IS and Sweet NJ. 1956. Gastrointestinal water and electrolytes. I. The equilibration of radiosodium in gastrointestinal contents and the proportion of exchangeable sodium in the gastrointestinal tract. *J. Clin. Invest.* 35: 502.

Edelman IS and Leibman J. 1959. Review: Anatomy of body water and electrolytes. *Am. J. Med.* 27: 256.

Foy JM and Schnieden H. 1960. Estimation of total body water (virtual tritium space) in the rat, cat, rabbit, guinea-pig and man, and of the biological half-life of tritium in man. *J. Physiol*. 154: 169–176.

Grollman A, Shapira AP and Gafford G. 1953. The volume of the extracellular fluid in experimental and human hypertension. *J. Clin. Invest*. 32(4): 312–316.

Guzel O, Olgim Erdikmen D, Ayolin D, Mutlu Z and Yildar E. 2012. Investigation of the effects of CO_2 insufflation on blood gas volumes during laparoscopic procedures in pigs. *Turk. J. Vet. Anim. Sci*. 36(2): 183–187.

Hardy RM and Osborne CA. 1979. Water deprivation test in the dog: Maximal normal values. *J. Am. Vet. Med. Assoc*. 174: 479.

Harrison JB, Sussman HH and Pickering DE. 1960. Fluid and electrolyte therapy in small animals. *J. Am. Vet. Med. Assoc*. 137: 637.

Haskins SC. 1984. Fluid and electrolyte therapy. *Compend. Contin. Educ. Pract. Vet*. 6: 244.

Hawk CT, Leary SL and Morris TH. 2005. In: *Formulary for Laboratory Animals*. Ames, Iowa: Blackwell.

Hobbs TR, O'Malley JP, Khouangsathiene S and Dubay CJ. 2010. Comparison of lactate, base excess, bicarbonate and pH as predictors of mortality after severe trauma in rhesus macaques (*Macaca mulatta*). *Comp. Med*. 60(3): 233–239.

Jorgensen KD, Ellegaard L, Klastrup S and Svensen O. 1998. Haematological and clinical chemical values in pregnant and juvenile Göttingen minipigs. *Scand. J. Lab. Anim. Sci*. 25: 181–190.

Kamis AB and Norrhayatea MN. 1981. Blood volume in Macaca fascicularis. *Primates* 22(2): 281–282.

Kovacikova J, Winter C, Loffing-Cueni D, Loffing J, et al. 2006. The connecting tubule is the main site of the furosemide-induced acidification by the vacuolar H^+-ATPase. *Kidney International* 70: 1706–1716.

Lee HB and Blaufox MD. 1985. Blood volume in the rat. *J. Nucl. Med*. 26(1): 72–76.

Lewis LD and Morris ML. 1987. *Small Animal Clinical Nutrition*. Topeka, Kansas: Mark Morris Associates.

Maxwell MH, Kleeman CR and Narins RG. 1987. *Clinical Disorders of Fluid and Electrolyte Metabolism*. New York: McGraw-Hill.

Muir WW and DiBartola SP. 1983. Fluid therapy. In: *Current Veterinary Therapy VIII*, Kirk RW (ed.), p. 28. Philadelphia: WB Saunders Co.

Nadell J, Sweet NJ and Edelman IS. 1956. Gastrointestinal water and electrolytes. II. The equilibration of radiopotassium in gastrointestinal contents and the proportion of exchangeable potassium in the gastrointestinal tract. *J. Clin. Invest*. 35: 512.

National Research Council. 2003. In: *Nutrient Requirements of Nonhuman Primates*, 2nd edition. ISBN 9780309069892.

O'Connor WJ and Potts DJ. 1969. The external water exchanges of normal laboratory dogs. *J. Exp. Physiol*. 54: 244.

O'Connor WJ and Potts DJ. 1988. Kidneys and drinking in dogs. In: *Renal Disease in Dogs and Cats: Comparative and Clinical Aspects*. Michell AR (ed.), p. 30. Oxford: Blackwell Scientific Publications.

Peters JP. 1948. The role of sodium in the production of oedema. *N. Engl. J. Med*. 239: 353.

Putnam DF. 1971. *Composition and Concentrative Properties of Human Urine*. Washington, DC: National Aeronautics and Space Administration.

Retzlaff JA, Newton Tauxe W, Kiely JM and Stroebel CF. 1969. Erythrocyte volume, plasma volume and lean body mass in adult men and women. *J. Haematology* 33(5): 649–667.

Riches AC, Sharp JG, Brynmor Thomas D and Vaugh Smith S. 1973. Blood volume determination in the mouse. *J. Physiol.* 228: 279–284.

Rivers JPW and Burger LH. 1989. Allometry in dog nutrition. In: *Nutrition of the Dog and Cat*, Waltham Symposium No.7, p. 67. Cambridge: Cambridge University Press.

Robertson GL. 1983. Thirst and vasopressin function in normal and disordered states of water balance. *J. Lab. Clin. Med.* 101: 351.

Robertson GL, Shelton RL and Athar S. 1976. The osmoregulation of vasopressin. *Kidney Int.* 10: 25.

Rose BD. 1984. *Clinical Physiology of Acid-Base and Electrolytes.* New York: McGraw-Hill.

Rose BD and Post TW. 1989. *Clinical Physiology of Acid-Base and Electrolyte Disorders*, p. 5. New York: McGraw-Hill.

Salas SP, Giacaman A and Via CP. 2004. Renal and hormonal effects of water deprivation in late-term pregnant rats. *Hypertension* 44: 334–339.

Sandeers-Beer BE, Spano YY, Golighty D, Lara A et al. 2011. Clinical monitoring and correlates of nephropathy in SIV-infected macaques during high-dose antiretroviral therapy. *AIDS Research and Therapy* 8:3: 1–11.

Sands JM, Ivy EJ and Beeuwkes R. 1985. Transmembrane potential of renal papillary epithelial cells: Effect of urea and DDAVP. *Am. J. Physiol.* 248: 762–766.

Schrier RW. 1988. Pathogenesis of sodium and water retention in high output and low output cardiac failure, nephrotic syndrome, cirrhosis and pregnancy. *N. Engl. J. Med.* 319: 1065.

Shaw MI, Beaulieu AD and Patience JF. 2006. Effect of diet composition on water consumption in growing pigs. *J. Anim. Sci.* 84: 3123–3132.

Sheng HP and Huggins RA. 1979. A review of body composition studies with emphasis on total body water and fat. *Am. J. Clin. Nutr.* 32(3): 630–647.

Starling EH. 1896. On the absorption of fluid from the connective tissue spaces. *J. Physiol.* 19: 312–326.

Suzuki MT, Hamano M, Cho F and Honjo S. 1989. Food and water intake, urinary and faecal output, and urinalysis in the wild-originated cynomolgus monkeys (*Macaca fascicularis*) under the indoor individually caged conditions. *Jikken Dobutsu* 38(1): 71–74.

Taggart R and Starr C. 1989. *The Unity and Diversity of Life.* p. 398. Belmont, Cal.: Wadsworth.

US EPA. 1988. *Recommendations for and Documentation of Biological Values for Use in Risk Assessment.* EPA/600/6-87/008.

van Vonderen IK, Kooistra HS and Fijnberk AD. 1997. Intra- and interindividual variation in urine osmolality and urine specific gravity in healthy pet dogs of various ages. *Journal of Veterinary Internal Medicine* 11(1): 30–35.

Walser M, Seldin DW and Grollman A. 1953. An evaluation of radiosulphate for the determination of the volume of extracellular fluid in man and dogs. *J. Clin. Invest.* 32: 299.

Webb RK, Woodhall PB, Tisher CC, Glaubiger G, Neelon FA and Robinson RR. 1977. Relationship between phosphaturia and acute hypercapnia in the rat. *J. Clin. Invest.* 60(4): 829–837.

Watson PE, Watson ID and Batt RD. 1980. Total body water volumes for adult males and females estimated from simple anthropometric measurements. *Am. J. Clin. Nutr.* 33(1): 27–39.

Weekley LB, Daldar A and Tapp E. 2003. Development of renal function tests for measurement of urine concentrating ability, urine acidification and glomerular filtration rate in female cynomolgus monkeys. *Contemporary Topics in Laboratory Animal Medicine* 42(3): 22.

Woodward KT Berman AR, Michaelson SM et al. 1968. Plasma erythrocyte and whole blood volume in the normal beagle. *Am. J. Vet. Res.* 29: 1935.

Zia-Amirhusseini P, Minthorn E, Benincosa LJ, Hart TK, Hottenstein CS, Tobia CAP and Davis CB. 1999. Pharmacokinetics and pharmacodynamics of SB-240563, a humanised monoclonal antibody directed to human interleukin-5 in monkeys. *J. Pharmacol. Exp. Thera.* 291(3): 1060–1067.

Zweens J, Frankena H and Zijlstra WG. 1978. The effect of pentobarbital anaesthesia upon the extracellular fluid volume in the dog: Study by continuous infusion and single injection models. *Pfugers Arch.* 376: 131.

chapter two

Physico-chemical factors

This chapter is a precursor to the details given later in this book concerning various ingredients of a sterile intravenous infusion solution. In modern preclinical safety assessment of novel pharmaceuticals the formulation of complex chemicals has become critical, particularly in intravenous infusion delivery systems. For both small and large (biopharmaceutical) molecules the resulting infusate very often is a complex mixture of active ingredient, solvent, solute, and buffering agents. Consequently, it is important to know the potential effects of all the individual ingredients so that appropriate controls can be employed and that possible effects attributable to the active ingredient can be properly assigned. This can be achieved only if these factors can be related to the natural physiological ability of circulating blood to compensate for the administration of xenobiotics by the vascular infusion route.

Introduction

The relationships between properties of xenobiotic formulations and blood will be explored. The most important physico-chemical properties are osmolality, pH, viscosity, surface tension, and diffusion, all of which are inherent and controlled properties of blood itself. We will see how it is important that vascular infusion formulations comply as closely as possible to the biological matrix (blood and its components) in order to achieve a successful infusion study. Similarly, in later chapters we will see how these factors are also important in the control of infusate volume and rate of delivery as well as compatibility with equipment materials and blood components.

Osmolality

Co-author: Dean Hatt

GlaxoSmithKline, UK

The effects of a solution after injection or infusion on the physiology of the vasculature of the subject have to be considered when determining the nature of that solution. The osmolality and tonicity of the solution are good indicators of the effects that may occur. Definitions of both parameters help us in the understanding of these terms.

The number of particles in a given amount of a fluid and their behaviour within the fluid are referred to as the **osmolality**. Blood is a typical example of such a fluid, with cellular components, large and small molecules, and charged ions in plasma water. Most solutions administered into the vascular system are also of similar composition, that is to say, a mixture of various large or small molecules, many carrying an ionic charge, within a simple or complex carrier fluid. Consequently, it can be deduced that it would be important for the osmolality of a formulation administered into general circulation to be compatible with the osmolality of blood itself.

The osmolality of any solute is dependent on the number of particles in the solution. One osmole (Osm) is defined as 1 gram molecular weight of any non-dissociable substance. If a compound in solution dissociates into two or three particles, the number of osmoles in solution is, correspondingly, increased two or three times. For example, assuming that NaCl completely dissociates in solution, each millimole of NaCl provides two milliosmoles (mOsm), which would be 1 mOsm of Na^+ and 1 mOsm of Cl^-. The milliosmolar concentration of a solution may be expressed as the milliosmolarity or milliosmolality of the solution. **Osmolality** refers to the number of osmoles per kilogram of solvent, that is, the concentration of osmotically 'active' particles in the solution, and is a function only of the number of particles and is not related to their molecular weight, size, shape, or charge. An aqueous solution with an osmolality of 1.0 results when 1 Osm of a solute is added to 1 kg of water. The volume of the resulting solution exceeds 1 L by the relatively small volume of the solute. **Osmolarity**, in contrast, refers to the number of osmoles (or osmotically active particles) per *litre* of solution. In biological fluids, there is a negligible difference between osmolality and osmolarity.

Serum or plasma osmolality may be measured by freezing point depression. One osmole of a solute in 1 kg of water depresses the freezing point of the water by 1.86°C (Smithline and Gardner 1976), and such a solution has an osmolality of 1 Osm/kg, or 1000 mOsm/kg. Average values for

Table 2.1 Measured Serum Osmolality Values

Species	Osmolality (mOsm/kg)	References
Human (60 kg)	275–298	Zarandona and Murdoch 2005
Rat (250 g)	342–354	Salas et al. 2004
Mouse (25 g)	310–325	Bernstein 1968
Dog (10 kg)	292–308	Hardy and Osborne 1979
Minipig (10 kg)	288	Thornton et al. 1989
NHP (2.5 kg)	267–320	Vrana et al. 2003

measured serum osmolality in laboratory animals and humans are given in Table 2.1.

A solution is said to be hyperosmotic if its osmolality is greater than that of plasma and hyposmotic if its osmolality is less than that of plasma. An isosmotic solution has an osmolality identical to that of plasma. In any fluid compartment, the **osmotic effect** of a solute is dependent on the permeability characteristics of the membranes surrounding the compartment. When the membranes are freely permeable to the solutes, allowing movement of the solutes in both directions down concentration gradients to achieve equilibrium on either side of the membrane, such solutes are termed **ineffective osmoles**, as they do not generate osmotic pressure. Conversely, when a molecule, such as glucose, cannot pass through the semi-permeable membrane, water, for example, will continue to pass down its concentration gradient, so diluting the **effective osmoles**, and this increase in fluid in the compartment will generate an osmotic pressure. Such osmolalities and osmotic pressures are major factors affecting the normal movement of fluids between the compartments.

The effective osmolality of a solution is also known as the **tonicity** of that solution. A measured osmolality of a solution would include both effective and ineffective osmoles, and thus the tonicity of a solution may be less than the measured osmolality if both effective and ineffective osmoles are present.

Changes in the osmolality of ECF may or may not initiate movement of water between intracellular and extracellular compartments. A change in the concentration of what are termed **permeant** solutes (e.g. urea, ethanol) does not cause movement of fluid, because these solutes are distributed equally throughout the total body water. On the other hand, a change in the concentration of **impermeant** solutes (e.g. glucose, sodium) does cause movement of fluid, because such solutes do not readily cross cell membranes. This is an important feature when considering the potential acceptability of an intravenous infusion formulation on the basis of haemocompatibility.

As can be seen in Table 2.1, plasma osmolality in laboratory animal species tends to be close to 300 mOsm/kg. Fluids with effective osmolalities greater than 300 mOsm/kg are hypertonic to plasma (the concentration of impermeant solutes is greater than plasma), and those less than 300 mOsm/kg are hypotonic to plasma (the concentration of impermeant solutes is less than plasma). Those with effective osmolalities of 300 mOsm/kg are said to be isotonic with plasma (concentration of impermeant solutes are equal) and is the desired value when considering formulations for intravenous delivery. As discussed in Chapter 1, intravenous infusion delivery of solutes results in an additive effect on total body fluid and consequentially results in alterations in body fluid space volumes and tonicity. These alterations elicit homeostatic shifts of fluid between compartments in an attempt to maintain isotonicity in fluid spaces.

To summarise, the effect on the vasculature is likely to be closely related to the number of 'effective osmoles' or tonicity of a solution, and a measurement of osmolality gives us a numerical value to quantitatively assess this.

Measurement of osmolality

Osmolality is measured using an osmometer to measure one of the colligative properties: osmotic pressure (mm Hg), boiling point elevation (°C), vapour pressure (mm Hg), and freezing point depression (°C). All of these properties are affected to a similar degree by the number of dissolved solutes. So one mole of any substance dissolved in 1 kg of water will cause an osmotic pressure of 17000 mm/kg, a boiling point elevation of 0.52°C, a vapour pressure decrease of 0.3 mm/Hg, and a freezing point depression of –1.86°C, although it is more common to express this as an osmolality of 1000 mOsm/kg of water, or 1000 mOsm, for short.

Measuring osmotic pressure across a semi-permeable membrane with water on one side and the solution in question on the other side appears simple. However, as there is no such thing as perfectly semi-permeable membrane, small ions would always travel through and, therefore, be impossible to measure. Measuring small changes in boiling point is messy, and as the change in boiling point is small, it is also problematic.

This leaves us with vapour pressure and freezing point depression, both of which are extensively used, but their respective colligative property is also the key to their limitation, and this needs to be understood to fully understand the osmolality value that they provide.

Vapour pressure osmometers are commonly used, but because all components exert their own pressure, some liquids, such as ethanol, provide a pressure that gives a misleading result.

With osmometers that use freezing point depression, the sample is 'supercooled' down to a temperature lower than the actual freezing point, and this value is compared to a calibration line created using two or more standard salt solutions with known freezing points that have been checked with a further 'isotonic' salt solution lying within the calibrated range. With solutions containing anything but low levels of solvents, the sample may fail to freeze and therefore a result can be unobtainable.

Plasma osmolality can be estimated from the following equation (Shull 1978; Feldman and Rosenberg 1981):

$$\text{Calculated plasma osmolality} = 2\text{Na}^+ + \frac{\text{BUN}}{2.8} + \frac{\text{Glucose}}{18}$$

where BUN is blood urea nitrogen.

In this equation, the concentrations of urea and glucose in milligrams per decilitre are converted to millimoles per litre by the conversion factors 2.8 and 18, respectively. The measured osmolality should not exceed the calculated osmolality by more than 10 mOsm/kg, as, in such a case, an abnormal osmolality gap is said to be present. This occurs when an unmeasured solute, as in the presence of an intravenous infusate, is present in large quantity.

Prediction of osmolality

Osmolality can be calculated mathematically (see Table 2.2) or predicted using known linear relationships of the solution's constituent parts. Figure 2.1 illustrates this linear relationship for three common solvents used in infusion formulations (1–10% DMSO, 1–10% ethanol, and 1–10% PEG400) analysed by both the vapour pressure and freezing point depression methods. In the case of DMSO and PEG400 the two methods of assessment yield very similar results, whereas the freezing point depression method for ethanol is largely unobtainable because of its extremely low freezing point.

Therefore, if the researcher knows the osmolality of the component parts, the osmolality of the mixture can be calculated. This is illustrated in Table 2.3.

From the data in Table 2.3,

$$(141.96/562.5) \times 1000 = 252 \text{ mOsm/L}$$

It follows that since aqueous solutes behave linearly (even in combination), the osmolality of a particular concentration can be predicted from higher or lower concentrations of its component parts, as is shown in Table 2.4.

Table 2.2 Calculated Plasma Osmolality Values[a]

Species	BUN (mg/dL)	Glucose (mg/dL)	Calculated osmolality (mOsm/kg)
Human (60 kg)	14.0 ± 3.6[b]	88.2 ± 7.2[b]	290
Rat (250 g)	15.7 ± 2.7	118.9 ± 16.3	298
Mouse (25 g)	15.4 ± 5.0	122.4 ± 24.4	310
Dog (10 kg)	13.6 ± 3.2	100.5 ± 9.9	307
Minipig (10 kg)	12.3 ± 3.1[b]	99.0 ± 30.6[b]	300
NHP (2.5 kg)	17.3 ± 3.2[c]	80.6 ± 9.9[c]	311

[a] Data from various preclinical contract research organisations.
[b] Chen et al. 2011.
[c] Aonidet et al. 1990.

From the data in Table 2.4, the osmolality of a 10% 2-hydroxy-beta-cyclodextrin in 0.8% (w/v) aqueous sodium chloride solution could be predicted by

$$((64/8) \times 10) + ((285/0.9) \times 0.8) = 400 \text{ mOsm/L}$$

Using this relationship, a formulation can be developed or modified easily to create a solution that is isotonic. If we know we need to use 8% 2-hydroxy-beta-cyclodextrin to solubilise our drug substance and we wish to target isotonicity (e.g. 292 mOsm/L), then we can calculate the percent of sodium chloride (X) required:

$$292 - 64 = ((285/0.9) \times X)$$

$$\text{Thus, } 228/X = 285/0.9 \text{ and } X = 0.72\%$$

Therefore, 8% 2-hydroxy-beta-cyclodextrin in 0.72% (w/v) aqueous sodium chloride is isotonic.

As a further extension of this, the osmolality of organic solvents and the drug substance can also be used in this way, as shown in Table 2.5.

From the data in Table 2.5, the osmolality of a 25 mg/mL formulation of Drug A in 5:95 v/v DMSO:0.9% w/v aqueous sodium chloride can be predicted. As we know the osmolality of our stock solution (50 mg/mL Drug A in 0.9% w/v aqueous sodium chloride), by subtracting the osmolality of vehicle (0.9% w/v aqueous sodium chloride):

$$347 - 285 \text{ (osmolality of 0.9\% w/v aqueous sodium chloride)} = 62 \text{ mOsm/L}$$

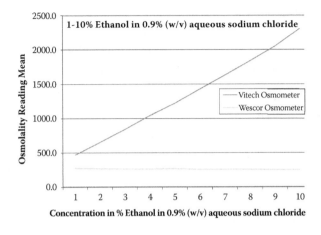

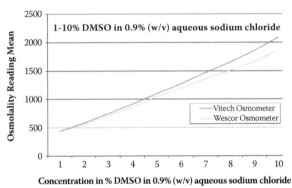

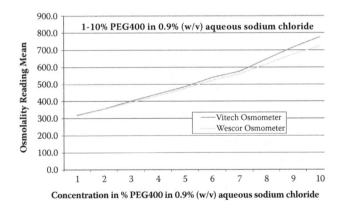

Figure 2.1 (See colour insert.) Mean osmolality reading for 1–10% DMSO, ethanol, and PEG 400 in 0.9% w/v aqueous sodium chloride measured by vapour pressure/freezing point depression.

Table 2.3 Example Estimation of the Osmolality of a Mixture

Constituent	Volume (mL)	mOsm/mL	mOsm
Sterile water for injection	500 mL	0	0
Sodium bicarbonate 8.4%	50 mL	2	100
Potassium chloride	10 mL	4	40
Heparin 5000 units	0.5 mL	0.46	0.23
Pyridoxine	1 mL	1.11	1.11
Thiamine	1 mL	0.62	0.62
Total	562.5 mL		141.96

Table 2.4 Osmolality of Three Typical Vehicle Components

Solution	Concentration	Osmolality (mOsm/L)
Aqueous sodium chloride	0.9%	285
Aqueous 2H beta cyclodextrin	8%	64
2H beta cyclodextrin in aqueous sodium chloride	8%/0.9%	349

Table 2.5 Osmolality of Sample Drug Formulation Components

Formulation ingredient	Concentration	mOsm/L
DMSO in water	1%	140
Drug A in 0.9% (w/v) aqueous sodium chloride	50 mg/mL	347

We have the osmolality of 50 mg/mL of Drug A and therefore the osmolality of 25 mg/mL of Drug A:

$$62 \times 25/50 = 31 \text{ mOsm/L}$$

As the formulation will contain 5% DMSO, the true osmolality of this component will be

$$5 \times 140 = 700 \text{ mOsm/L}$$

As the formulation will contain 95% (0.9% (w/v) aqueous sodium chloride), the osmolality of this component will be

$$285 \times 95/100 = 271 \text{ mOsm/L}$$

Using all of this, the osmolality of a 25 mg/mL formulation of Drug A in 5:95 v/v DMSO:0.9% w/v aqueous sodium chloride should approximate to

$$31 + 700 + 271 = 1002 \text{ mOsm/L}$$

Note: In the above example, the displacement of the drug has not been taken into consideration, so the actual quantity of vehicle and therefore osmolality may be slightly less at high drug concentrations.

However, importantly when dealing with organic solvents, the osmolality measurements may be variable, and with freezing point depression the appearance of high solvent content may prevent freezing of the sample and preclude any measurements.

Furthermore, whilst increasingly higher concentrations of drugs in solution do demonstrate a linear relationship, the molecular weight (MW) of a drug in question appears to have no relationship with osmolality, as the same concentration of two drugs of similar MW in the same vehicle can have completely different osmolality values. It is therefore not possible to predict osmolality using MW or concentration without a previous measurement.

The true effect of osmolality

As discussed in the first part of this chapter, the true effect of osmolality is the tonicity of the solution. The effects of changes in tonicity on red blood cells are demonstrated in Figure 2.2. It can be seen from the illustrations in this figure that, with a hypertonic solution, fluid is pulled out of the cells into the vasculature space, causing crenation (shrinking) of the red blood cells. With a hypotonic solution, fluid moves into the cells from the vasculature space, causing cell rupture (haemolysis) of the red blood cells. With an isotonic solution, there is an equal movement of fluid into and out of the red blood cells and the cell volume remains constant.

Although this simple relationship between tonicity and the effect on red blood cells is well understood, the actual effect on red blood cells goes beyond that of tonicity, as the total dose volume is equally if not more important. The number of osmotically effective particles in contact with the red blood cells is higher with a greater dose volume. An example of acceptable osmolality values versus dose volume is shown in Table 2.6.

Another important factor to be considered is the rate of infusion, as the number of osmotically effective particles in contact with the red blood cells is higher with a slower infusion rate. However, unlike dose volume,

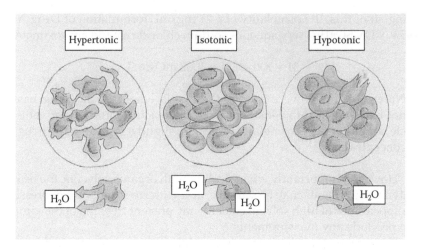

Figure 2.2 (See colour insert.) Effects of changes in tonicity on a solution of red blood cells.

Table 2.6 Predicted Acceptable Osmolality Range for Specific Dose Volumes

Dose volume (mL/kg)	Acceptable osmolality range (mOsm/kg)
20	275–325
10	250–350
5	200–400
1	0–800[#]

[#] In reality, whilst an osmolality of 0 would have little effect on red blood cells at a dose volume of 1 mL/kg, in practise, addition of solutes to raise the osmolality is easy and therefore preferred.

it is not clear whether an acceptable osmolality range is proportionate to the infusion rate, and this may be dependent on the solution itself. What is clear is that a slow infusion of a similar volume of a solvent with or without drug will likely be better tolerated than a fast one. However, the action of this solution 'trickling' into a vein will be more likely to result in precipitation of drug and therefore infusion line blockage, not to mention the greater potential for pH and osmolality effects as the solution mixes with the blood. It is, therefore, important that careful consideration be given when choosing the dose volume and infusion rate, and this consideration should include osmolality.

There are other factors where the osmolality (measured effect) does not necessarily lead to the expected tonicity (true effect). Below are three clear cases that need to be taken into account.

Solutions containing solvents

As many solvents have a major effect on the colligative property being measured, such as freezing point, the osmolality values are often extremely high and/or variable. A good example of this is a formulation containing DMSO:

2% (v/v) DMSO in sterile water – 299 mOsm/kg

It would be expected that this formulation behaves as isotonic and would result in no overall net movement of fluid from red blood cells. However, intravenous administration of this formulation would, in fact, cause severe haemolysis. Consider another formulation containing DMSO:

5% (v/v) DMSO in saline – 1094 mOsm/kg

In this case, the formulation would be expected to behave as hypertonic, with movement of fluid out of the cell. However, intravenous administration of this formulation would, in fact, result in no net movement of fluid. In essence, with many solvents, such as DMSO and ethanol, the real effect of the solution would be that of the vehicle without the solvent component. In the examples above, the effect of the solution in which DMSO sits, such as sterile water or saline, is the deciding factor. It may therefore be appropriate to measure or predict the osmolality of just the aqueous components. However, it should be known that at higher solvent concentrations, the solvent may have a direct irritant effect on red blood cells.

Solutions containing high drug concentrations

As the osmolality of a solution is a measure of the number of solutes, it is not surprising that a high concentration of drug in solution will have a high osmolality. Consider two osmotically inactive drugs in saline solutions:

50 mg/mL Drug A in saline – 1745 mOsm/kg
50 mg/mL Drug B in saline – 1850 mOsm/kg

These osmolality values would be expected to behave as hypertonic, with movement of fluid out of the cell. Here, in the case of Drug A, the solution behaves as if it is isotonic, whereas the Drug B solution behaves as if hypotonic. The reason for these apparent anomalies lies with the concept of 'effective' osmolality. Neither Drug A nor Drug B is osmotically active itself and, therefore, their real effect is governed by the effect of the vehicle, in this case saline, which is isotonic, thereby resulting in the Drug A solution having no effect on red blood cells. With Drug B, although the solution is effectively isotonic and the drug not osmotically active, Drug B has a direct irritant effect on red blood cells that leads to haemolysis and is, therefore, suggestive of a hypotonic solution.

Solutions containing glucose

Whilst it is not possible to reduce the osmolality of a hypertonic solution once prepared, although re-formulation with less osmotically active components should be considered, hypotonic solutions can be manipulated after preparation by addition of solutes. This typically involves addition of either sodium chloride or glucose, as both demonstrate good linearity with respect to osmolality:

10% cyclodextrin in sterile water – 83 mOsm/kg
10% cyclodextrin in sterile water
　+ 0.7% sodium chloride – 301 mOsm/kg
10% cyclodextrin in sterile water
　+ 4% glucose – 299 mOsm/kg

Here, we would expect the addition of sodium chloride or of glucose to behave as if isotonic, and certainly the addition of sodium chloride does. However, the addition of glucose can make the resulting solution behave differently. After administration, glucose is rapidly metabolised, leaving the other components behind. Consequently, the true effect of the solution may be hypotonic, resulting in fluid entering the red blood cells and leading to haemolysis.

In all three cases discussed here, the osmolality does not result in the expected effect on red blood cells. This demonstrates that the true effect of osmolality, osmotically active effect, or tonicity, whilst not measurable in a quantitative sense, should be used along with the osmolality figures to predict a likely outcome and is one that demonstrates the need for blood compatibility testing.

Acid-base balance

An acid is a proton donor and a base a proton acceptor. In the following equation, HA is an acid and A⁻ a base:

$$HA \leftrightarrow H^+ + A^-$$

The acidity of a solution refers to the chemical activity of its constituent H^+ ions. The concentration of H^+ ions in body fluids is in the range of nano-equivalents per litre, which is considerably lower than other important electrolytes. Hydrogen ions are highly reactive. The proteins of the body have many dissociable groups that may gain or shed protons as the concentration of H^+ changes, resulting in alterations in charge and molecular configuration that may affect protein structure and function. The concentration of H^+ in body fluids must be kept constant so that detrimental changes in enzyme function and cellular structure do not occur, hence the need to buffer infusate formulations sufficiently to be compatible with the blood matrix. To maintain cellular function, the body of mammals has elaborate mechanisms that maintain blood H^+ concentration within a narrow range, typically 37 to 43 nmol/L (pH 7.43 to 7.37, where pH = $-\log[H^+]$) and ideally 40 nmol/L (pH = 7.40). Therefore, blood is normally slightly alkaline; below pH 7.35 it is considered too acidic and above pH 7.45 too alkaline (Porter and Kaplan 2011; Waugh and Grant 2007). As discussed later in detail, blood pH, partial pressure of oxygen (pO_2), partial pressure of carbon dioxide (pCO_2), and HCO_3^- are carefully regulated by a number of homeostatic mechanisms that exert their influence principally through the respiratory and urinary systems in order to control the acid-base balance. Hence the normal blood pH is kept relatively constant. As the blood has its own complex buffering mechanisms (see below) to achieve this consistency, it is possible for the pH of xenobiotic infusions to vary somewhat from the pH 7.40 normal level. The kidneys help remove excess chemicals from the blood, and they ultimately remove H^+ ions and other components of the pH buffers that build up in excess. The acidosis that would result from failure of the kidneys to perform this function is known as *metabolic acidosis*. However, excretion by the kidneys is a relatively slow process and may take too long to prevent acute acidosis resulting from a sudden decrease in pH, such as during exercise. Under these circumstances the lungs provide a faster way to control the pH of blood. The increased breathing response to exercise helps to counteract the pH-lowering effects of exercise by removing CO_2, a component of the principal pH buffer in the blood. Acidosis that results from the failure of the lungs to eliminate CO_2 as fast as it is produced is known as *respiratory acidosis*.

The term *pH* is defined as the negative base 10 logarithm of the hydrogen ion concentration [H^+] expressed in equivalents per litre:

$$pH = -\log_{10}[H^+] = \log_{10}(1/[H^+])$$

Thus, at the normal extracellular fluid [H^+] of 40 nEq/L (4 x 10^{-8} Eq/L):

$$pH = -\log_{10}(4 \times 10^{-8})$$
$$= -\log_{10}4 - \log_{10}10^{-8}$$
$$= -(0.602)-(-8)$$
$$= 7.398$$

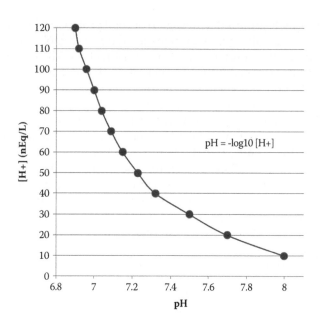

Figure 2.3 Exponential relationship between [H⁺] and pH.

There is an inverse relationship between pH and [H⁺]: the greater the [H⁺], the lower the pH. The relationship is exponential, as shown in Figure 2.3.

Buffering

A buffer is a compound that can accept or donate protons (hydrogen ions) and minimise a change in pH. A buffer solution consists of a weak acid and its conjugate salt. When a strong acid is added to a buffer solution containing a weaker acid and its salt, the dissociated protons from the strong acid are donated to the salt of the weak acid and the change in pH is minimised. This can be represented by the Henderson–Hasselbalch equation:

$$pH = pK_a + \log \frac{[\text{salt}]}{[\text{acid}]}$$

where pK_a is the dissociation constant.

It is common practice, therefore, that, when preparing formulation solutions for intravenous delivery, the formulation is buffered to as close to normal blood pH (7.4) as possible by adding acids or bases such as hydrochloric acid (HCl) or sodium hydroxide (NaOH). If the amount of strong acid (e.g. HCl) or base (e.g. NaOH) added to a solution of a weak

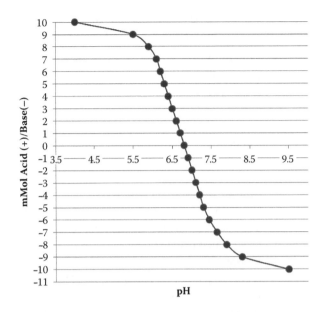

Figure 2.4 Titration curve for an aqueous solution containing a buffer.

acid and its salt (i.e. a buffer solution) is plotted against pH, the resulting relationship produces a sigmoidal curve (Figure 2.4).

In the pH range associated with the greatest slope of the curve, the change in pH is smallest for a given amount of added acid or base, and buffer capacity is greatest at the midpoint of the curve. At this point, there are equal amounts of weak acid and its conjugate salt and, in accordance with the Henderson–Hasselbalch equation, pH will equal pK_a. The region of best buffer capacity extends approximately 1.0 pH unit on either side of the pK_a. Thus a buffer is most effective within one pH unit of its pK_a, and compounds with pK_a values in the range 6.4–8.4 are considered the most useful as buffers in biologic systems. The pK_a of some important biological compounds are listed in Table 2.7.

Of course, the body has its own buffering capacity. These can be divided into *bicarbonate*, which is the primary buffer system of extracellular fluid, and *non-bicarbonate* buffers (e.g. proteins and inorganic and organic phosphates), which constitute the primary intracellular buffer system.

The bicarbonate–carbonic acid system

In a vascular infusion system we are primarily concerned with the extra-cellular component and buffering capacity within it, hence the interest in this bicarbonate–carbonic acid system.

Table 2.7 pK$_a$ Values of Biologically
Important Compounds

Compound	pK$_a$
Phosphoric acid	2.0
Citric acid	2.9
Carbonic acid	3.6
Lactic acid	3.9
Citrate^{1-}	4.3
Acetic acid	4.6
Creatinine	5.0
Citrate^{2-}	5.6
Uric acid	5.8
Organic phosphates	6.0–7.5
Oxygenated haemoglobin	6.7
Phosphate^{1-}	6.8
α-Amino (amino terminal)	7.4–7.9
Deoxygenated haemoglobin	7.9
Ammonium	9.2
Bicarbonate	9.8
Phosphate^{2-}	12.4

Gaseous CO_2 produced in the tissues is soluble in water, and the concentration of dissolved CO_2 in body fluids is proportional to the partial pressure of CO_2 in the gaseous phase (Pco$_2$):

$$[CO_{2diss}] = \alpha(Pco_2)$$

where α is a factor called the *solubility coefficient of CO_2*.

The solubility coefficient of CO_2 has a value of 0.0301 mmol/L/mm in arterial plasma at 37°C. Thus,

$$[CO_{2diss}] = 0.0301Pco_2$$

Dissolved CO_2 combines with water to form carbonic acid:

$$CO_{2diss} + H_2O \rightarrow H_2CO_3$$

This reaction proceeds slowly without catalysis but is dramatically increased by the enzyme carbonic anhydrase, which is present abundantly in the body in red blood cells. In the body, therefore, the hydration of CO_2 to form H_2CO_3 reaches equilibrium almost instantaneously (Malnic and

Giebisch 1972). Through a series of equations this can be represented by the Henderson–Hasselbalch equation:

$$pH = pK_a + \log \frac{[HCO_3^-]}{[CO_{2diss}]}$$

At blood pH of 7.4, the value of pK_a is 6.1, and applying the solubility coefficient for CO_2,

$$pH = 6.1 + \log \frac{[HCO_3^-]}{0.03 P_{CO_2}}$$

This is the clinically relevant form of the equation and shows that, in body fluids, pH is a function of the ratio between HCO_3^- concentration and P_{CO_2}.

It is considered that one of the most important factors to the success of this natural buffering system is that the bicarbonate–carbonic acid buffer pair functions as an open system. In such an open system, carbonic acid, in the presence of carbonic anhydrase, forms CO_2, which is eliminated entirely from the system by alveolar ventilation.

$$CO_{2diss} + H_2O \leftrightarrow H_2CO_3 \leftrightarrow H^+ + HCO_3^-$$

The acid member of the buffer pair is free to change directly with the salt member as compensation for occurring metabolic acidosis. If P_{CO_2} is kept constant at 40 mm Hg, the effectiveness of the bicarbonate–carbonic acid system is increased significantly. However, normally the body reduces P_{CO_2} below the normal value of 40 mm Hg, thus increasing the effectiveness further. This open system, together with P_{CO_2} closely regulated by alveolar ventilation, is a demonstration of how effective the body is in regulating circulatory blood pH, particularly in a pending acidosis scenario. Consequently, it is often considered that the natural buffering capacity of blood would more easily deal with more acidic (low pH) formulations than alkali (high pH) formulations.

Proteins

Plasma proteins play a limited role in extracellular buffering, whereas intracellular proteins play an important role in the total buffer response of the body. The buffer effect of proteins is due to their dissociable side groups. For most proteins, including haemoglobin, the most important of these dissociable groups is the imidazole ring of histidine residues (pK_a 6.4–7.0). Amino acid terminal amino groups (pK_a 7.4–7.9) also contribute to the buffer effect of proteins. Haemoglobin is responsible for over 80% of the non-bicarbonate buffering capacity of whole blood, whereas plasma proteins contribute 20%. Of the plasma proteins, albumin is much more important than the globulins (van Styke et al. 1920; van Leeuwen 1964; Madias and Cohen 1982).

Phosphates

The most important intracellular buffers are proteins and inorganic and organic phosphates. The pK_a value of $H_2PO_4^-$ is 6.8, and pK_a values for organic phosphates range from 6.0 to 7.5. Inorganic phosphate is a more important buffer intracellularly, where its concentration is high (approximately 40 mEq/L in skeletal muscle cells), and less important in extracellular fluid, where its concentration is much lower (approximately 2 mEq/L).

Acidosis and alkalosis

In conditions of vascular infusion of xenobiotic formulations, there is potential to induce an *acidosis* or *alkalosis*, depending on the pH and pK_a of the formulation in question. These conditions are probably the biggest risk from vascular infusates, and therefore it is important to understand what these terms mean. Acidosis and alkalosis refer to the pathophysiological processes that cause net accumulation of acid or alkali in the body. The terms *acidaemia* and *alkalaemia* refer specifically to the pH of extracellular fluid. In acidaemia the extracellular fluid pH is lower than normal and the [H+] higher than normal. In alkalaemia the extracellular fluid pH is higher than normal and the [H+] lower than normal. This is important to understand since in conditions of respiratory alkalosis, blood pH can be within the normal range because of effective renal compensation: the condition of alkalosis is present, but not alkalaemia. Also, there can be *mixed* acid-base disturbances in which blood pH remains within normal range as a consequence of a combination of counterbalancing acid-base disturbances. Should these circumstances arise during a vascular infusion of a xenobiotic formulation, potential clinical effects could be observed, such as changes in respiratory rate and depth as well as changes in plasma electrolytes and urine production.

There are four primary acid-base disturbances: *metabolic* and *respiratory acidosis*, and *metabolic* and *respiratory alkalosis*. The metabolic disturbances refer to net excess or deficit of non-volatile acid, whereas the respiratory disturbances refer to the net excess or deficit of volatile acid (dissolved CO_2). Metabolic acidosis is characterised by a decrease in plasma HCO_3^- concentration and a decrease in pH (increased [H+]) caused by either HCO_3^- loss or buffering of a non-volatile acid. Metabolic alkalosis is characterised by an increase in plasma HCO_3^- concentration and increased pH (decreased [H+]), usually as a result of loss of chloride ions and fluid or low levels of the weak acid albumin (hypoalbuminaemia). It is more difficult to produce a metabolic alkalosis by administration of an alkaline solution. The respiratory counterparts are characterised by increased Pco_2 (acidosis/hypercapnia caused by alveolar hypoventilation) or decreased Pco_2 (alkalosis/hypocapnia caused by hyperventilation). The most common response likely in a vascular infusion scenario is that of metabolic acid-base disturbances that would have the respiratory counterparts as a compensatory response.

As shown in Table 2.8, each primary metabolic or respiratory acid-base disturbance is accompanied by a secondary change in the opposing component of the system. The compensatory response involves the component opposite the one disturbed and moves the pH of the system toward but not completely back to normal.

In conclusion, when administering heavily buffered formulations by vascular infusion it would be advisable to measure blood gases at intervals during the delivery term in order to evaluate the potential cause of any consequent acid-base disturbances; objective physical findings suggestive of an acid-base disturbance, such as hyperventilation, would be unreliable indicators in isolation. Care needs to be taken in blood sample collection and handling when analysing for blood gases. Arterial samples are preferred to venous ones because oxygenation of blood can also be measured and the sample would not be affected by stasis of blood flow and local tissue metabolism. The difference between arterial and venous blood will obviously be the difference in Po_2, which demonstrates the oxygenation of the blood in the lungs and utilisation in the tissues. Conversely, arterial samples may not reflect the acid-base status in peripheral tissues. The Pco_2 is slightly higher and the pH slightly lower in venous samples because of local tissue metabolism (Tables 2.9 and 2.10). Normal blood gas values for laboratory animals should be established by the laboratory performing the analysis in order to normalise the potential effects of the sample collection and handling methods.

The importance of neutralising the pH of infusate solutions is, therefore, demonstrated since small changes in $[H^+]$ either side of physiological

Table 2.8 Primary and Compensatory Acid-Base Disturbances

Acid-base disturbance	pH	$[H^+]$	Primary consequence	Compensatory respiratory response
Metabolic acidosis	↓	↑	↓$[HCO_3^-]$	↓Pco_2
Metabolic alkalosis	↑	↓	↑$[HCO_3^-]$	↑Pco_2
Respiratory acidosis	↓	↑	↑$[HCO_3^-]$	↑Pco_2
Respiratory alkalosis	↑	↓	↓$[HCO_3^-]$	↓Pco_2

(Rose 1989)

Table 2.9 Normal Arterial Blood Gas Values in Laboratory Animals

Species	pH	Pco$_2$ (mm Hg)	HCO$_3^-$ (mEq/L)	Po$_2$ (mm Hg)	References
Human	7.34–7.44	35–45	22–26	75–100	University of Texas Southwestern Medical Center, Dallas, USA; Koul et al. 2011
Rat	7.12–7.5	35–40	19–39	65–460	Pakulla 2004; Grant and McGrath 1986
Mouse	7.46 ± 0.2	39 ± 3	29 ± 4	88 ± 3	Lee et al. 2009
Dog	7.41 ± 0.03	36.8 ± 2.7	21.8 ± 1.5	97.1 ± 7.7	Haskins 1983; Ilkiw et al. 1991
Minipig	7.5	40 ± 3	31.2 ± 0.9	182 ± 22	*Ellegaard Newsletter* 34, 2010
NHP	7.36 ± 0.16	27–35	26.7–29.2	90–95	Joseph and Morton 1971; Binns et al. 1972

Table 2.10 Normal Jugular Venous Blood Gas Values in Laboratory Animals

Species	pH	Pco$_2$ (mm Hg)	HCO$_3^-$ mEq/L	Po$_2$ (mm Hg)	References
Human	7.32–7.46	40–50	27.4–30.2	30–40	Chu et al. 2003; Williams 1998
Rat	7.25–7.36	48.2–58.1	24.3–31.5	33.5–42.1	Goundasheva 2000
Mouse	7.23–7.32	nda	nda	nda	
Dog	7.35 ± 0.02	42.1 ± 4.4	22.1 ± 2.0	55.0 ± 9.6	Ilkiw et al. 1991
Minipig	7.32 ± 0.06	50.3–64.4	27.7–31.4	134.9–181.2	Guzel et al. 2012
NHP	7.39 ± 0.09	29.7 ± 2.4	24.3 ± 5.2	96.3 ± 4.9	Hobbs et al. 2010; Taguchi et al. 2012

nda: no data available

blood pH can elicit acidosis or alkalosis, which may confound the interpretation of similar responses attributable to an adverse effect of the
active ingredient. Therefore, in cases where the infusate is known to be
at variance with physiological blood pH of the species, measurement
of blood gases routinely would be an advantage to assist in the correct

interpretation of resultant side effects. Acid-base disturbances would be considered when abnormalities in total CO_2 or electrolytes (Na^+, K^+, Cl^-) are observed in the biochemical profile. The CO_2 concentration may be increased as a consequence of metabolic alkalosis or a renal adaptation to respiratory acidosis. Total CO_2 may be decreased as a consequence of metabolic acidosis or a renal adaptation to respiratory alkalosis. Therefore, the acid-base disturbance cannot be determined on the basis of the total CO_2 concentration alone. Objective clinical findings suggestive of an acid-base disturbance, such as hyperventilation, are unreliable as indicators and may not be evident in some species. Blood gas analysis is required to identify and classify acid-base disorders conclusively.

In suspected incidences of acid-base disturbance in vascular infusion studies, the blood pH should be first evaluated. Evaluation of pH often provides the answer to the question of whether or not an acid-base disturbance is present. If the pH is outside the normal range (pH 7.43 to 7.37), an acid-base disturbance is present. If the pH is within the normal range, an acid-base disturbance may or may not be present. If the subject is acidaemic and plasma HCO_3^- concentration is decreased, metabolic acidosis is present. If the subject is acidaemic and Pco_2 is increased, respiratory acidosis is present. If the subject is alkalaemic and plasma HCO_3^- concentration is increased, metabolic alkalosis is present. If the subject is alkalaemic and Pco_2 is decreased, respiratory alkalosis is present (Harrington et al. 1982). These relationships are represented in Figure 2.5.

Secondly, it would be useful to calculate the expected compensatory response in the opposing component of the system—for example, respiratory alkalosis as compensation for metabolic acidosis or metabolic alkalosis as compensation for respiratory acidosis (see Table 2.8). Depending on the magnitude of the change in the adaptive or secondary response, a simple case of mixed acid-base disorder would be suspected. In the case of mixed disorder the magnitude in the change of blood pH can be significant. For example, the effect on extracellular pH in a mixed disorder circumstance is minimised if Pco_2 and HCO_3^- change in the same direction (e.g. respiratory acidosis and metabolic alkalosis) and is maximised if the disorders change Pco_2 and HCO_3^- in opposite directions (e.g. respiratory acidosis and metabolic acidosis). In the case of the former, blood pH may even remain within the normal range, whereas in the latter case blood pH would be markedly abnormal.

Routinely, the approach to acid-base evaluation focuses on the relationship between pH, HCO_3^-, and Pco_2 as described by the Henderson–Hasselbalch equation, in which pH is shown to be a function of HCO_3^- concentration and Pco_2. The Pco_2 is viewed as the respiratory component and is determined by alveolar ventilation, whereas the HCO_3^- concentration is considered the metabolic (or nonrespiratory) component and is regulated by the kidneys. However, only Pco_2 is independent, such

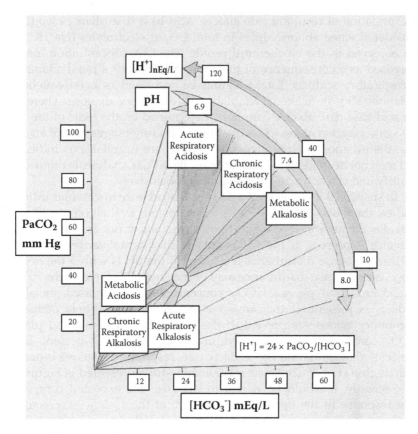

Figure 2.5 (See colour insert.) Acid-base map.

that when a primary increase in Pco_2 occurs, the hydrogen ions that are produced by dissociation of H_2CO_3 are buffered by proteins (notably haemoglobin) and the HCO_3^- concentration increases secondarily. Interestingly, the control or consequences of acid-base disturbances are more complicated than this and also involve other potential biomarkers of change, most notably the circulating anions and cations. In the 1980s it was proposed by Stewart and co-workers (Stewart 1981, 1983) that acid-base chemistry in biological systems is significantly influenced by maintenance of electroneutrality and dissociation equilibria of incompletely dissociated solutes. These variables coincide with the potential chemical makeup of essential infusate vehicles and solutes, and it is proposed that any reported differences in the extracellular concentrations of strong ions would be indicative of acid-base disturbance. This concept is known as the *strong ion difference* (SID), and in fact the potential of acid-base disturbance following vascular infusion can be determined by the assessment of SID and the total concentration of weak acid and/or P_{CO_2}. The SID changes if the difference between the sum of strong cations and the sum of strong anions changes. Ions are considered strong if they are almost completely dissociated at the pH of body fluids. Of the strong cations (Na^+, K^+, Ca^{2+}, and Mg^+) only sodium is really at high enough concentration in extracellular fluid that a change in its concentration is likely to have a substantial effect on SID. Similarly, of the strong anions (Cl^-, lactate, SO_4^-, and PO_4^-) chloride is the only strong anion routinely measured that could have a substantial effect on SID. The normal ranges of plasma cations and anions are given in Table 2.11.

There have been many variations on the concepts of using measured anions and cations to determine acid-base disturbances, and this could be a useful marker to include in vascular infusion studies with novel agents in order to monitor the potential for such disturbances (Fencl and Rossing 1989; Fencl and Leith 1993; Figge et al. 1992). Of course, the assessment of SID can be adjusted in terms of accuracy depending on what measurements are made. The ranges of common strong cations and strong anions have been mentioned previously, with the exception of plasma proteins. Automated clinical chemistry analysers provide values for serum/plasma sodium, potassium, chloride, and proteins, and these ions are normally part of routine clinical chemistry assessments. Of course, measurement of these ions alone do not obey the laws of electroneutrality, largely because of unmeasured cations and anions, and the difference is commonly known as the *anion gap*. It would be important to establish the extent of the anion gap for each animal using the routine measurements of serum/plasma ions prior to the commencement of infusion so that deviations from 'normal' can be routinely monitored to give an indication of the extent of potential acid-base disturbance as a consequence of the infusate. The anion gap usually implies the serum anion gap, but the urine anion gap is also a clinically useful measure (Kirschbaum et al. 1999). The anion

Table 2.11 Anion Gap from Routine Clinical Chemistries in Laboratory Animals

	Electrolyte concentration (mEq/L)						
Species	Na^+	K^+	Total cations	Cl^-	HCO_3^-	Total anions	Anion gap (mEq/L)
Human	140	4.3	144.3	101	26	127	+17.3
Rat	143	3.7	146.7	104	29	133	+13.7
Mouse	149	5.5	154.5	109	29	138	+16.5
Dog	148	4.3	152.3	110	22	132	+20.3
Minipig	145	6.4	151.4	100	30	130	+21.4
NHP	150	4.4	154.4	106	27	133	+21.4

gap is an artificial and calculated measurement that is representative of the unmeasured ions in plasma or serum. Commonly measured cations include sodium (Na^+), potassium (K^+), calcium (Ca^+), and magnesium (Mg^+). Cations that are generally considered 'unmeasured' include a few normally occurring serum proteins. Likewise, commonly 'measured' anions include chloride (Cl^-), bicarbonate (HCO_3^-), and phosphate ($H_2PO_4^-$), while commonly 'unmeasured' anions include sulphates and a number of serum proteins. By definition, only Na^+, K^+, Cl^-, and HCO_3^- are used when calculating the anion gap. In normal health there are more measurable cations compared to measurable anions in the serum/plasma; therefore, the anion gap is usually positive.

Whilst the measurement and assessment of the anion gap can be indicative of the presence of a circulating acid-base disturbance, the cause will be assumed to be a result of the infusate within the controls on such a study. There is also the potential that any xenobiotic under evaluation could have some effect on respiratory or renal function, which may itself confound any assessment of acid-base disturbances in such studies. In either event the shift in anion gap can be useful in the identification of mixed acid-base disturbances (de Morais 1992).

It should be remembered that the biochemical profile can provide useful information about the acid-base status of the patient (de Morais and Muir 1995). First, the total CO_2 should be used to estimate $[HCO_3^-]$. Unfortunately the respiratory component cannot be assessed using the biochemical profile; hence, decreases in total CO_2 related to respiratory alkalosis or increases in total CO_2 related to respiratory acidosis cannot be predicted. The second step is to look at the strong (Na^+, Cl^-) and weak (albumin, phosphate) ions that affect $[HCO_3^-]$. Total CO_2 will increase (i.e. $[HCO_3^-]$ will increase) with decreases in chloride (hypochloraemic alkalosis), increases in sodium (contraction or concentration alkalosis), or decreases in albumin (hypoalbuminaemic alkalosis) concentrations. Total CO_2 will decrease (i.e. $[HCO_3^-]$ will decrease) with increases in chloride (hyperchloraemic acidosis), decreases in sodium (expansion or dilutional acidosis), increases in unidentified strong anion (organic acidosis), or increases in phosphate (hyperphosphataemic acidosis) concentrations. Unmeasured strong anions cannot be accurately evaluated using the biochemical profile, but they can be estimated from the anion gap. An increase in anion gap in the absence of hyperphosphataemia strongly suggests the presence of an organic acidosis. Chloride disorders can be identified easily whenever the sodium concentration is normal by simply evaluating the $[Na^+]-[Cl^-]$ difference. In this setting, an increase in $[Na^+]-[Cl^-]$ suggests the presence of metabolic alkalosis, whereas a decrease in $[Na^+]-[Cl^-]$ suggests the presence of a hyperchloraemic acidosis.

A complete description of whole-body regulation of acid-base balance is beyond the scope of this publication and is well documented, but in general terms it requires the cooperation of the liver, the kidneys, and the lungs. By the process of alveolar ventilation, the lungs remove a large amount of volatile acid produced each day by metabolic processes. The liver metabolises amino acids derived from protein catabolism to glucose or triglyceride and releases NH_4^+ in the process. When urea is synthesised in the liver from NH_4^+ and CO_2, H^+ is produced and HCO_3^- is titrated. Consequently, the liver produces much of the fixed or non-volatile acid that must be excreted each day. The kidneys excrete NH_4^+ in the urine, thus diverting it from ureagenesis and producing a net gain of HCO_3^- and net loss of H^+. All these processes are very adaptable to various influences, whether as a consequence of disease or from the introduction of a xenobiotic or other non-physiological event, including vascular infusion into normovolaemic models.

Viscosity

Viscosity is a consequence of 'shear stress', which is more associated with solids, where it is the consequence of forces being applied to fixed bodies, potentially resulting in the distortion of solid bodies that are fixed in only one plane. This particular property of a fluid, be it blood itself or a complex infusate, can be accommodated within the biological system by various physiological adaptations, such as force of flow and expansion/contraction of blood vessels. Consequently, it is considered that the viscosity of an infusate would most likely affect its flow through the delivery system itself, which has defined dimensions and capacities. These laws of shear stress can also be applied to liquids. However, a liquid is unable to support a shear stress but flows under its action. Consider the steady flow of a liquid as represented in Figure 2.6.

The liquid may be considered to consist of layers of very thin sheets or laminas, and the flowing motion, maintained by the shear stress, consists of the relative sliding motion of these laminas. The liquid in contact with the two sides (top and bottom) of this two-dimensional schematic is at rest, and the velocity of any lamina increases with its distance from the sides, being fastest in the centre. The ratio $\Delta v / \Delta y$ for adjacent laminas is the rate of change of velocity for the lamina with the distance from the nearest side, or dv/dy. Since no permanent strain is possible for a liquid, Hooke's law is not directly applicable, but pure liquids would follow the law that can be represented as follows:

$$\text{Shear stress in liquid} = \eta \frac{dv}{dy}$$

where η is a constant known as the *viscosity* of the liquid.

The units of shear stress are $N\,m^{-2}$, and the units of the velocity gradient dv/dy are s^{-1}. Hence the units of η are $N\,m^{-2}\,s$, or $kg\,m^{-1}\,s^{-1}$.

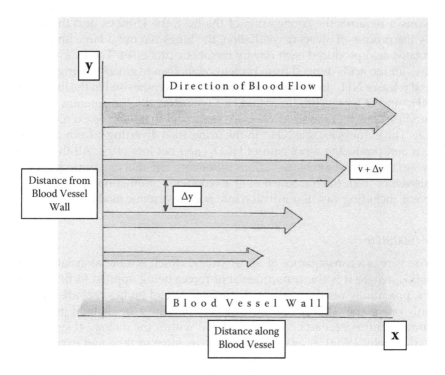

Figure 2.6 (See colour insert.) Blood flow within a blood vessel.

Viscosity is a kind of internal liquid friction, retarding the flow of fluids through tubes and retarding the motion of objects through the fluid. The flow of liquids through very narrow cylindrical tubes was first investigated by Jean Louis Poiseuille in 1846, and his work in relation to the flow of blood in veins and arteries and the capillary network in general has been summarised (Landis 1933). Poiseuille showed that the volume of a liquid of viscosity η flowing per second through a tube of internal radius a when a pressure difference $p_1 - p_2$ existed between its ends was given by

$$\text{Rate of flow} = \frac{\pi a^4}{8\eta} \frac{(p_1 - p_2)}{L}$$

where L is the length of the tube.

The difference of pressure between the ends divided by the length of the tube is known as the *pressure gradient*, and Poiseuille's result shows that the rate of flow is proportional to this pressure gradient. This formula has an immediate application to the choice of needle size in a hypodermic syringe. Since the radius of the tube occurs to the fourth power, needle size is much more important than the pushing pressure in determining the outflow from a syringe. Doubling the diameter of the needle has the same effect as increasing the push force 16 times.

Although Poiseuille's formula holds well for pure liquids, it does not hold for suspensions or dispersions, mixtures of different kinds of material. Blood is an anomalous liquid of this type. For blood, doubling the pressure difference between the ends of the tube does not simply double the flow: it may be much higher than this. In other words, the viscosity of the blood itself decreases as the shear stress in it increases. A liquid that obeys Poiseuille's law of flow is called a Newtonian liquid; blood is a non-Newtonian liquid.

In fact, viscosity coefficients can be defined in two ways:

- **Dynamic viscosity**, also known as *absolute viscosity*. The SI physical unit is the pascal-second (Pa.s), and the cgs physical unit is the poise (P), named after Jean Louis Marie Poiseuille, more commonly expressed in centipoises (cP).
- **Kinematic viscosity** is the dynamic viscosity divided by the density (typical units are cm^2/s and the stokes).

Viscosity is generally independent of pressure and tends to fall as temperature increases (for example, water viscosity goes from 1.79 cP to 0.28 cP in the temperature range from 0°C to 100°C). For many vascular infusion formulations, it would be a very desirable ability to predict the viscosity of a blend of two or more liquids. Whilst the mathematical steps are beyond the scope of this book, this can be estimated using the Refutas equation (Maples 2000).

The reciprocal of viscosity is **fluidity**, usually symbolised by φ or F, depending on the convention, and is measured in *reciprocal poise* (cm.s.g^{-1}), sometimes called the *rhe*. Fluidity is a parameter that is seldom used these days.

It is important to understand the natural physiological effects of viscosity of the blood on flow and pressure, since the introduction of a xenobiotic infusion into the capillary network will disrupt this system. Clearly the viscosity of the infusate will potentially have an effect both within the delivery system itself as well as within the internal blood capillary network. Consequently, viscosity is a measurement that should be known for all infusion formulations and should be kept within controllable limits (see Table 2.12).

In many vascular infusion situations researchers are concerned with the ratio of the inertial force (movement of fluid or 'flow') to the viscous force (forces restricting flow), which takes into consideration all of these physical properties of fluids. It is a mathematical attempt to produce one number for the relationship between these forces, the *Reynolds number*.

Table 2.12 Physical Properties of Some Liquids at Room Temperature (unless otherwise stated)

Substance	Density (p) (kg m⁻³)	Dynamic viscosity (η) Pa.s	cP	Surface tension (S) (10^{-3} N m⁻¹)
Blood (37°C)	1075	0.003500	3.500	53
Water	1000	0.001002	1.002	73
Sodium chloride	1013	0.001003	1.003	83
Ethanol	789	0.001074	1.074	22
Ethylene glycol	1110	0.016100	16.100	48
Glycerol	1260	1.200000	1200.000	63

Such calculations were initially designed for engineering situations back in the nineteenth century (Stokes 1851; Reynolds 1883). A Reynolds number can be determined for a number of different situations where a fluid is in relative motion to a surface. These definitions generally include the fluid properties of density and viscosity, plus a velocity and a characteristic length or characteristic dimension of the boundaries of the fluid (blood vessel or catheter and extension line). The mathematical relationship is represented as follows:

$$Re = \frac{\rho v L}{\mu} = \frac{vL}{v}$$

Where
  v = mean velocity of the object relative to the fluid (m/s)
  L = characteristic linear dimension (travelled length of the fluid) (m)
  μ = dynamic viscosity of the fluid (Pa.s or N.s/m or kg/m.s)
  v = kinematic viscosity (m²/s)
  ρ = density of the fluid (kg/m³)

Surface tension

One of the most important properties of a liquid is the tendency for its surface to contract. The surface behaves like an elastic skin that constantly tries to decrease its area so that, consequently, globules of the liquid are formed that would be as near spherical as possible. The tension in the surface of a liquid is independent of the area and is called the *surface tension*, defined as the force per unit length acting across any line drawn in the surface and tending to pull the surface apart across the line.

  Surface tension arises because the molecules near the surface are closer together than those deep inside the liquid, and this implies a certain surface energy. The surface tension S of a liquid can be regarded as the potential energy per unit area of the surface. The units of S are either N m^{-1} or J m^{-2}, which are of course equivalent, and some typical values are given in Table 2.12. The tendency of a liquid to assume a configuration of minimum surface area is a consequence of the general principle of conservation of energy. Surface energy, or surface tension, is a mutual property of two materials that share a common surface.

Whilst unless indicated to the contrary, surface tension of a liquid can be assumed to be relative to air. However, infusion delivery systems are essentially closed systems with no exposure to air. But, in this regard, a very important consequence of the existence of surface tension is that there exists a difference of pressure across any curved surface separating two fluids. The pressure on the concave side exceeds the pressure on the convex side by an amount, which can be represented thus:

$$p_1 - p_2 = 2S/r$$

where r is the radius of the surface.

This relationship may be used to determine the rise of a liquid in a capillary tube, as shown in Figure 2.7.

When viewing the capillary tube system, the system will equilibrate when the pressure at points 1 and 2 are equal, since these two points are on the same level. But the pressure p_2 at point 2 is greater than the pressure p_3 at point 3 just inside the upper curved surface by an amount pgh, where p is the density of the liquid, and the pressure p_3 is less than the pressure p_4 at point 4 on the concave side of the upper surface by an amount $2S/r$. In this particular case, $p_1 = p_4 =$ atmospheric pressure. This scenario is typical of this open system, and the rise of fluid in the capillary tube will be proportional to atmospheric pressure. Rather than go into the complex maths relating to the relationships in this model, we should consider the added complexities presented by an infusion system, because under certain conditions it is expected that, in a similar fashion, blood from circulation will enter the implanted catheter. As mentioned previously, infusion systems are closed systems and are therefore not subjected to atmospheric pressure conditions. These are indeed more complex since there will be the pressure presented by the blood within the cannulated blood vessel as well as the pressure presented by the delivery device. The radius (r) of the catheter is clearly a factor in this equation and will obviously differ between the species being used for the study. It is likely that the pressure exerted by the pumping device at point 5 will be marginally greater than that at the tip of the catheter, point 6, and this will be dependent on the length and diameter of the total tubing, as well as on the durometry of the tubing, the capacity of the polymer material of the tubing to expand to accommodate fluids under pressure in the closed system. It will also be important to take into consideration the blood pressure within the

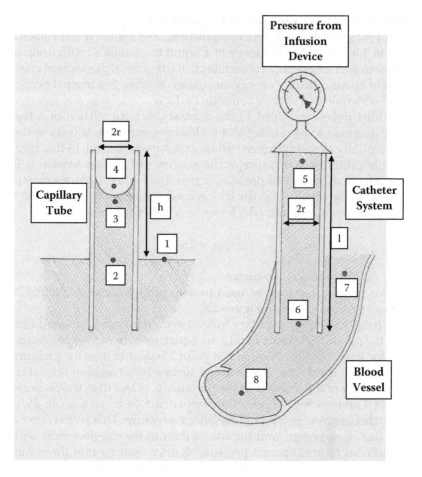

Figure 2.7 Comparison of the rise of fluid in a capillary tube with flow of fluid in a fine intravenous catheter.

blood vessel itself, both distal (point 7) and proximal (point 8) to the tip of the implanted catheter. Often in the cannulated vessels of small animals (rats and mice in particular) the distal side of the blood vessel, such as the femoral vein, is sacrificed; but as the tip of the catheter is then located in the posterior vena cava, there is natural blood flow under core venous return pressure past the implanted catheter. As a result of these variables as well as the potential viscosity and electrostatic effects of the formulation, it is important to run a compatibility trial of the final formulation with the entire infusion system under experimental conditions. Figure 2.8 illustrates the complexity of forces affecting the surface tension of fluids throughout an infusion system.

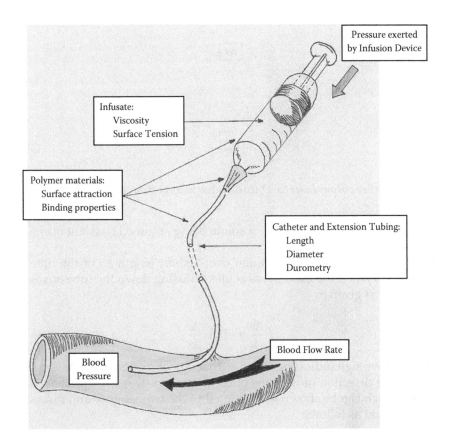

Figure 2.8 (See colour insert.) Forces affecting surface tension in an infusion system.

Diffusion

When considering the physical forces at work within a closed infusion system, it is difficult to leave out the concept of *diffusion*. Whilst it is a force of little consequence to vascular delivery outside of the system itself, it is a force that can have profound effects on dosing accuracy during very low volume and low rate infusions.

The diffusion of a solute within a system such as an infusion system tends to obey Fick's law, which states that the rate of diffusion per unit area in a direction perpendicular to the area is proportional to the gradient of concentration of solute in that direction. The concentration is the mass of solute per unit volume. Consider the example of a tube (blood vessel) of cross-sectional area A down which a solute is diffusing

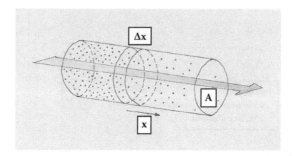

Figure 2.9 (See colour insert.) Diffusion due to a concentration gradient.

(Figure 2.9), the concentration of solute being assumed constant over any cross-section of the tube.

If the change of concentration over a short length Δx of the tube or vessel is Δc, then the mass Δm of solute diffusing down the tube or vessel in time Δt is given by

$$\frac{\Delta m}{\Delta t} = -DA\ \frac{\Delta c}{\Delta x},$$

the negative sign indicating that the direction of motion of solute is opposite to the direction of increase of concentration. D is the diffusion constant, which can be shown to be related to the temperature and viscosity of the liquid as follows:

$$D = kT/6\pi a \eta$$

where a is the radius of the particle of solute assumed spherical and k is Boltzmann's constant.

Consequently, the flow of an infusate through an infusion system into the blood stream has an inherent speed based on diffusion along a concentration gradient.

Summary of infusion forces

The acceptability of a solution for infusion should be based not only on the potential pharmacological and toxicological effects of its component parts, but also on the physiological properties that may affect the vasculature. These include pH and osmolality, and whilst the former is well understood, the latter needs considerable thought. Osmolality, whilst it is measurable and/or can be calculated or predicted, is only an indication of the tonicity (effective osmolality), and therefore the difference between osmolality and likely tonicity needs to be considered, alongside dose

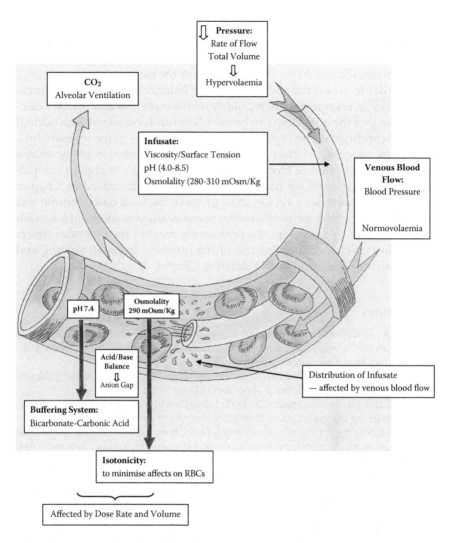

Figure 2.10 (See colour insert.) Schematic representation of bioregulatory forces during intravenous infusion delivery.

volume and infusion rate, in determining a dosing vehicle and regime (Figure 2.10).

In vascular infusion safety assessment studies it is relatively common practice to be measuring serum/plasma parameters and/or urinary parameters. Consequently, data are invariable available on such studies to evaluate the anion gap. Because we know that plasma is electro-neutral we can conclude that the anion gap calculation represents the concentration of unmeasured anions. The anion gap varies in response to changes in the concentrations of the aforementioned serum components that contribute to the acid-base balance. Therefore, calculating the anion gap is clinically useful, as it helps in the differential diagnosis of a number of disease states that could be a consequence of the infusate itself.

In order to assess the state of acid-base balance during vascular infusion studies, it is important to regularly monitor plasma and urinary electrolytes so that the anion gap can be calculated and compared with normal vascular conditions. The physico-chemical properties of the infusate (pH, osmolality, viscosity, surface tension, density) should also be measured and compared to that of blood in the relevant species to ensure compatibility. Pre-study tests for haemocompatibility are discussed in Chapter 5. In addition, consideration has to be given to the total daily volume and rate of delivery of the infusate in order to maintain physiological tolerability, as both these parameters can profoundly modify the potential effects of the physico-chemical properties of the infusate. Infusion volume and rate of delivery are discussed in detail in Chapter 3.

References

Aonidet A, Bouisson H, de la Farge F and Valdiguié P. 1990. Serum reference values of the cynomolgus monkey: A model for the study of atherosclerosis. *J. Clin. Chem. Clin. Biochem.* 28: 251–252.

Bernstein SE. 1968. Physiological characteristics. In: *Biology of the Laboratory Mouse*, 2nd edition, Green EL (ed.). New York: Dover Publications.

Binns R, Clark GC and Simpson CR. 1972. Lung function and blood gas characteristics in the rhesus monkey. *Laboratory Animals* 6: 189–198.

Chen Y, Qin S, Ding Y, Li S, Yang G, Zhang J, Li Y, Cheng J and Lu Y. 2011. Reference values of biochemical and haematological parameters for Guizhon minipigs. *Exp. Biol. Med.* 236(4): 477–482.

Chu YC, Chen CZ, Lee CH, Chen CW, Chang HY and Hsiue TR. 2003. Prediction of arterial blood gas values from venous blood gas values in patients with acute respiratory failure receiving mechanical ventilation. *J. Formos. Med. Assoc.* 102(8): 539–543.

De Morais HSA. 1992. A non-traditional approach to acid-base disorders. In: *Fluid Therapy in Small Animal Practice*, DiBartola SP (ed.), p. 297. Philadelphia: WB Saunders Co.

De Morais HSA and Muir WW. 1995. Strong ions and acid-base disorders. In: *Kirks's Current Veterinary Therapy XII*, Bonagura JD (ed.), p. 121. Philadelphia: WB Saunders Co.

Feldman BF and Rosenberg DP. 1981. Clinical use of anion and osmolal groups in veterinary medicine. *J. Am. Vet. Med. Assoc.* 178: 396.

Fencl V and Rossing TH. 1989. Acid-base disorders in critical care medicine. *Annu. Rev. Med.* 40: 17.

Fencl V and Leith DE. 1993. Stewart's quantitative acid-base chemistry: Applications in biology and medicine. *Respir. Physiol.* 91: 1.

Figge J, Mydosh T and Fencl V. 1992. Serum proteins and acid-base equilibria: A follow-up. *J. Lab.Clin. Med.* 120: 713.

Grant TL and McGrath JC. 1986. The effects of arterial blood gas tensions on pressor responses to angiotensin II in the pithed rat. *J. Physiol.* 372: 437–444.

Goundasheva D. 2000. Effect of exercise on acid-base and blood gas changes in rats challenged with an inflammation agent. *Revue Méd. Vét.* 151(11): 1041–1046.

Guzel O, Erdikmen DO, Aydin D, Mutlu Z and Yildar E. 2012. Investigation of the effects of CO_2 insufflation on blood gas values during laparoscopic procedures in pigs. *Turk. J. Vet. Anim. Sci.* 36(2): 183–187.

Hardy RM and Osborne OA. 1979. Water deprivation test in the dog: Maximal normal values. *J. Am. Vet. Med. Assoc.* 174: 479.

Harrington JT, Cohen JJ and Kassirer JP. 1982. Introduction to the clinical acid-base disturbances. In: *Acid-Base*, Cohen JJ and Kassirer JP (eds), p. 119. Boston: Little, Brown and Co.

Haskins SC. 1983. Blood gases and acid-base balance: Clinical interpretation and therapeutic implications. In: *Current Veterinary Therapy VIII*, Kirk RW (ed.), p. 201. Philadelphia: WB Saunders Co.

Hobbs TR, O'Malley JP, Khouangsathiene S and Dubay CJ. 2010. Comparison of lactate, base excess, bicarbonate and pH as predictors of mortality after severe trauma in rhesus macaques (*Macaca mulatta*). *Comp. Med.* 60(3): 233–239.

Ilkiw JE, Rose RJ and Martin ICA. 1991. A comparison of simultaneously collected arterial, mixed venous, jugular venous and cephalic venous blood samples in the assessment of blood gas and acid-base status in dogs. *J. Vet. Intern. Med.* 5: 294.

Joseph WL and Morton DL. 1971. Long-term survival in the immediately functioning transplanted primate lung: Physiological findings. *Ann. Thoracic Surg.* 11: 442–449.

Kirschbaum B, Sica D and Anderson FP. 1999. Urine electrolytes and the urine anion and osmolar gaps. *J. Lab. and Clin. Med.* 133(6): 597–604.

Koul PA, Khan KH, Wani AA, Eachkoti R et al. 2011. Comparison and agreement between venous and arterial gas analysis in cardiopulmonary patients in Kashmir Valley of the India subcontinent. *Ann. Thoracic Med.* 6(1): 33–37.

Landis EM. 1933. Poiseuille's law and the capillary circulation. *Am. J. Physiol.* 103(2): 432–443.

Lee EJ, Woodske ME, Zou B and O'Donnell CP. 2009. Dynamic blood gas analysis in conscious, unrestrained C57BL/6J mice during exposure to intermittent hypoxia. *J. Appl. Physiol.* 107: 290–294.

Madias NE and Cohen JJ. 1982. Acid-base chemistry and buffering. In: *Acid-Base*, Cohen JJ and Kassirer JP (eds), pp. 13–17. Boston: Little, Brown and Co.

Malnic G and Giebisch G. 1972. Mechanism of renal hydrogen ion secretion. *Kidney Int.* 1: 280.

Maples RE. 2000. *Petroleum Refinery Process Economics*, 2nd edition. Tulsa, Oklahoma: PennWell Books.

Pakulla MA, Obal D and Loer SA. 2004. Continuous intra-arterial blood gas monitoring in rats. *Laboratory Animals* 38: 133–137.

Porter RS and Kaplan JL. 2011. Acid-base regulation and disorders. In: *Merck Manual of Diagnosis and Therapy*, professional edition. Merck & Co. Inc.

Reynolds O. 1883. An experimental investigation of the circumstances which determine whether the motion of water shall be direct or sinuous, and of the law of resistance in parallel channels. *Philosophical Transactions of the Royal Society* 174: 935–982.

Rose BD. 1989. *Clinical Physiology of Acid-Base and Electrolyte Disorders*, 3rd edition, p. 269. New York: McGraw-Hill.

Salas SP, Giacaman A and Vio CP. 2004. Renal and hormonal effects of water deprivation in late-term pregnant rats. *Hypertension* 44(3): 334–339.

Shull RM. 1978. The value of anion gap and osmolal gap determinations in veterinary medicine. *Vet. Clin. Pathol.* 7: 12.

Smithline N and Gardner KD. 1976. Gaps: Anionic and osmolal. *JAMA* 236: 1594.

Stewart PA. 1981. *How to Understand Acid-Base*. New York: Elsevier.

Stewart PA. 1983. Modern quantitative acid-base chemistry. *Can. J. Physiol. Pharmacol.* 61: 1444.

Stokes G. 1851. On the effect of the internal friction of fluids on the motion of pendulums. *Transactions of the Cambridge Philosophical Society* 9: 8–106.

Taguchi K, Watanabe H, Sakai H, Horinouchi H, Kobayashi K, Maruyama T and Otagiri M. 2012. A fourteen-day observation and pharmacokinetic evaluation after a massive intravenous infusion of hemoglobin vesicles (artificial oxygen carriers) in cynomolgus monkeys. *J. Drug Metab. Toxicol.* 3(4): 1–7.

Thornton SN, Forshing ML and Baldwin BA. 1989. Drinking and vasopressin release following central injections of angiotensin II in minipigs. *Quarterly J. Exp. Physiol.* 74: 211–214.

Van Leeuwen AM. 1964. Net cation equivalency ('base binding power') of the plasma proteins: A study of ion-protein interaction in human plasma by means of *in vivo* ultrafiltration and equilibrium dialysis. *Acta. Med. Scand.* 422(1): 1.

Van Styke DD, Hastings AB, Hiller A et al. 1920. Studies of gas and electrolyte equilibria in blood XIV: The amounts of alkali bound by serum albumin and globulin. *J. Biol. Chem.* 79: 769.

Vrana KE, Hipp JD, Goss AM, McCool BA, Riddle DR, Walker SJ, Wettstein PJ. et al. 2003. Nonhuman primate parthenogenetic stem cells. *PNAS* 100(1): 11911–11916.

Waugh A and Grant A. 2007. In: *Anatomy and Physiology in Health and Illness*, 10th edition, p. 22. Churchill Livingstone Elsevier.

Williams AJ. 1998. ABC of oxygen: Assessing and interpreting arterial blood gases and acid-base balance. *BMJ* 317: 1213–1216.

Zarandona M and Murdoch G. 2005. Sudden onset of polydipsia and polyuria. *Neuropathy*. Published online.

chapter three

Vascular infusion dynamics

'Vascular Infusion Dynamics' is a phrase used to describe the forces directly involved in the delivery of a liquid formulation onto the central or peripheral circulatory system. Simply, the dynamics involve the rate of delivery of the infusate, the total volume over a given period, and the blood flow/pressure within the delivery vessel. Of course, there are other factors that may influence these dynamics, many of which relate to the properties of the infusate and blood itself, as described in other chapters. There is also the intervention of the delivery catheter to be considered. Consequently, with an understanding of the movement of fluid normally throughout the body, its intake and elimination, and the forces involved in the maintenance of homeostasis, it is possible to use variables such as flow rates and infusion volumes to successfully achieve the delivery of complex and challenging formulations that may be on the very boundaries of physiological acceptability. Similarly, these factors under direct control of the infusionist can also negate, or at least reduce, fluid homeostatic issues resulting from the fluid loading of normovolaemic animal models.

Essential physiology

When thinking about the effects of implanted catheters present in the blood stream in venous blood vessels, we should also consider the specific dynamics of the flow of blood occurring naturally in the blood vessels in which the infusion is taking place. There is very little information available in published literature concerning the flow rates within the major venous vessels used for infusion in the various laboratory animal species. Therefore, blood pressure has been chosen as the marker for relative blood flows in these vessels (Table 3.1).

Table 3.1 Localised Blood Pressures in Laboratory Animals

| Species | Blood pressure (mmHg) measurement in: | | | | |
	Central venous pressure	Peripheral (non-invasive)	Carotid artery	Femoral artery	References
Human	5.0	126 ± 30	nda	135 ± 31	Ochsner et al. 1951; Mignini et al. 2006
Rat	0.5 ± 0.8	128 ± 5	137 ± 1	130 ± 5	Tanaka et al. 2005; Bunag and Butterfield 1982
Mouse	0.8 ± 0.5	123 ± 2	122 ± 2	116 ± 1	Krege et al. 1995; Mattson 1998; Scheuermann-Freestone et al. 2001
Dog	1–5	152 ± 38	141 ± 23	154 ± 31	Chalifoux et al. 1985; Nelson et al. 2010
Minipig	2–<8	140 ± 7	135 ± 5	116 ± 6	Scheepe et al. 2007; Myrie et al. 2006; Guo et al. 2011
NHP	nda	nda	nda	nda	
Rabbit	13–20	142 ± 4	122 ± 12	169 ± 16	Xu et al. 1998; Kumiko and Yoshiaki 1990; Hoefer et al. 2001; Herrold et al. 1992

nda: no data available.

During the delivery of a xenobiotic to central or peripheral circulation, the blood flow in the region will play a major part in the dissemination of the infusate from the point of delivery, dependent on the circumstances of blood flowing past the tip of the implanted catheter. In femoral and jugular vein cannulations, particularly in small animals such as rats and mice, it is routine to 'sacrifice', or ligate, the distal portion of the blood vessel as a part of the catheter implantation procedure. This practice results in the loss of flow of blood past the catheter tip within these blood vessels. This situation is thought to increase the possibility of blood clot formation at the tip of the catheter, which can be a life-threatening situation or at least enhance the incidence of some routine pathologies such as pulmonary granuloma formation (see Volume II). Therefore, it is now considered best practice in both femoral and jugular vein cannulations to advance the implanted catheter slightly beyond the blood vessel where the initial phlebotomy was made. For example, in small mammals (rats, mice, and rabbits) a jugular vein catheter would be advanced such that the tip would sit in the flow of the anterior vena cava, and the cephalic end of the jugular vein distal to the phlebotomy would be ligated and sutures used to anchor the catheter in place. The catheter itself would normally occupy the full lumen of the jugular vein in these small animals. Similarly, in femoral vein cannulations, the catheter tip would be advanced into the posterior vena cava up to the level of the renal vein branches, and the femoral vein distal to the phlebotomy would be ligated. Consequently, in both cases there is 'normal' venous return flow and pressure past the tip of the catheter. This is the catheter placement technique preferred by most researchers for the cannulation of these blood vessels for infusion delivery. The dynamics of fluid delivery via such placed catheters into circulating blood is complex and is influenced by the physico-chemical properties of the infusate discussed in previous sections of this book.

The data in Table 3.1 show the variable ranges for blood pressure measurements in the most common arterial regions, made principally by peripheral non-invasive means (often involving assessment of pulse strength in peripheral vessels using an inflatable limb cuff). Such data for arterial systolic blood pressure are readily available for most of the main laboratory animal species and humans. However, blood pressure within the venous system is less commonly available, particularly in the specific blood vessels often used for venous infusion delivery of xenobiotics. Similarly, data on actual blood flow rates in the specific areas of the venous system, such as the posterior and anterior vena cava and the jugular and femoral veins, are very difficult to find. In order to optimise the intravenous delivery of a xenobiotic, it is important that the flow of the infusate from a venous vascular infusion system be similar to that of the blood within the cannulated vessel. By matching up the rate of delivery (or pressure of delivery) in this way, the potential for effects on dissolution of the xenobiotic and the acute pressure effects of the resultant hypervolaemia will be minimised.

For Table 3.1, sufficient data could not be found for the measurement of venous return in the commonly used vessels (jugular and femoral veins, and anterior and posterior vena cava) of various laboratory animal species. Consequently, an indication of the central venous pressure is given for the various species. In fact, it can be seen that, within a wide range of values, there are significant similarities of arterial systolic pressure and central venous pressures across all the species.

Under the circumstances described for vascular cannulations, it is important to consider the potential range of flow rates and volumes that could be used for xenobiotic formulation deliveries that would not adversely affect the normal physiological blood flow regulation in the animal model. The remainder of this chapter focuses on these parameters in relation to intravenous delivery of xenobiotics.

Intravenous delivery rates and volumes

Literature review

In the recent past there have been a small number of publications attempting to give some guidance on the acceptability of fluid volumes and rates of delivery by a number of routes of administration in laboratory animals. Specifically for intravenous administration, these range considerably, from suggested maximum rates and volumes based on estimated circulating blood volumes, to more comprehensive reviews. In a published document from the Joint Working Group on Refinement (2001) it is suggested that, generally, maximum volumes should typically be 4% and no more than 5% of circulating blood volume. Also, it is stated quite arbitrarily that intravenous infusion volumes should be based on kidney function data, which generally translated into 5% circulating blood volumes per hour or 4 mL/kg body weight/hour. Additionally, in a publication called *The Care and Feeding of an IACUC* (Podolsky and Lukas 1999), produced for the training and benefit of preclinical researchers in the United States, it is suggested, for intravenous bolus dosing, that 10.0 mL/kg body weight is a reasonable guideline for all laboratory animal species with the exception of mice, for which 20.0 mL/kg body weight is suggested.

As a consequence of these variations, researchers in the preclinical field of this technology normally make reference to the more detailed reviews of this subject by Diehl et al. (2001) and Morton et al. (1997) when determining acceptable start-points for their vascular infusion studies. In the earlier publication it was recognised by the authors that, for rats, repeated doses of up to 40 mL/kg body weight at a rate of 1 mL/min (about 3 to 4 mL/min per kg) were commonly used without clinical or biochemical evidence of adverse effects. The study carried out by Morton and co-workers explored the potential for doubling this volume of isotonic saline solution infusion to 80 mL/kg body weight delivered at the same rate of 1 mL/min, once daily for 4 days. It is also noted that this study was performed in restrained rats with the administration route via the tail vein. This route is generally accepted as not being conducive to the administration of large volumes in short periods (as discussed earlier). Not surprisingly perhaps, it was concluded that this high rate of intravenous infusion of isotonic saline was essentially without adverse effects related to excess intravascular fluid volume. A number of changes relating to tachypnea and transient pulmonary oedema can be related to the restraining technique, and it is postulated that higher volumes and rates of delivery could possibly be achieved in a more ambulatory model, provided that the rate of infusion does not exceed the ability of the body to handle the transiently increased intravascular volume and to distribute and excrete the excess fluid volume. One potential measure of the maximal ability to excrete excess fluid is the glomerular filtration rate. In rats the glomerular filtration rate has been estimated at between 8.7 and 11.5 mL/min per kg body weight (Bivin et al. 1979; Clausen and Tyssebotn 1973). The infusion rates used by Morton et al. (1997) were much lower than this, suggesting the potential for infusing much larger volumes is possible.

In the later review by Diehl et al. (2001) a more comprehensive review of potential maximum intravenous infusion volumes and rates are presented for all the laboratory species. In that publication the proposed maximum infusion volumes and rates of delivery are very conservative and do not reflect any relationship with the mean circulating blood volumes, which are also presented in the publication for the purposes of determining potential maximum blood withdrawal volumes for the various species.

Therefore, because of these somewhat conflicting data, the following section attempts to bring all these data together in order to provide the researcher with some clear limits and ranges of infusion volumes and rates in order to achieve successful intravenous infusion deliveries and keep the physiological sequelae to a minimum.

Guidance for consideration

Of course there will be a wide range of potential infusion rates and volumes for use in non-clinical animal models; this would be necessary in order to achieve dose levels of xenobiotics that would provide a range of responses in order to properly define toxicological responses and no-effect (safe) levels. Therefore, delivery rates and volumes will range from low levels to very high levels. Invariably at the low end of the scale the lowest rates are determined by the limitations of the equipment. However, many modern infusion pumps can operate with considerable accuracy at rates of microlitres per hour; consequently, for infusions in this range there would be no adverse physiological issues of concern. Of course, it would be necessary at such low rates and small delivery volumes to devise a protocol of delivery that would negate the potential for loss of catheter patency, mainly a result of the presence of blood in the system and thrombus formation. At the other end of the scale, the limitations to high delivery rates and large volumes are generally physiological in nature and what could be acceptable from an animal welfare standpoint. It is this latter situation that the rest of this section will concentrate on.

Within the preclinical safety assessment community there is strict regulation concerning the animal welfare conditions when using infusion technology of any description, as evidenced in the UK with the Animals (Scientific Procedures) Act 1986. For the purpose of satisfying these requirements in the UK, an attempt was made in the late 1990s to define acceptable limits for total volumes and infusion rates for intravenous delivery to rodents and non-rodents. The ranges of rates of delivery and total volumes were based upon isotonic 0.9% sodium chloride formulations, which would be as neutral as possible with the normal physiology of body fluid homeostasis of the animal model being used. The concept was to build a profile of venous delivery values based on known and accepted volumes, from bolus administration to 24 hour/day infusion delivery, that were already considered to be physiologically acceptable and of no detriment to the animal model. Tables 3.2 and 3.3 give the figures that were generally considered acceptable at the time for rodent and non-rodent species.

Table 3.2 Rates and Volumes of Intravenous Infusion Delivery – Rodents

				Duration of administration			
1–2 min	3–10 min	For each minute up to 1 hour[a]	Up to 1 hour[1]	For each hour up to 6 hours[b]	Up to 6 hours[1]	For each hour up to 24 hours[c]	Up to 24 hours or longer[1]
			Maximum dose volume (mL/kg)[2]				
5(10)	10(15)	+0.2	25	+2	35	+2.5	80
	Rate (mL/kg/min)				**Rate (mL/kg/hour)**		
2.5(10)	1(5)	-	0.42	-	5.8	-	3.3

[1] Volumes specified indicate maximum (total) dose volume (mL/kg) for specified time period.
[2] Wherever possible the dose volume should be kept to the minimum practical.

Examples of how to use the data in Table 3.2 are as follows:

a A 30-minute infusion would use a maximum dose volume of 15 + (20 × 0.2) = 19 mL/kg, where the dose volume for up to 10 minutes is 15 mL/kg, and the dose volume for the additional 20 minutes is 0.2 mL/kg/min.

b A 4-hour infusion could use a maximum dose volume of 25 + (3 × 2) = 31 mL/kg, where the dose volume for up to 1 hour is 25 mL/kg, and the dose volume for the additional 3 hours is 2 mL/kg/hour.

c An 8-hour infusion could use a maximum dose volume of 35 + (2 × 2.5) = 40 mL/kg, where the dose volume for up to 6 hours is 35 mL/kg, and the dose volume for the additional 2 hours is 2.5 mL/kg/hour.

Similar data have been generated for non-rodent species in Table 3.3, and similar examples can be followed as for the rodent data in Table 3.2.

In the preparation of these two tables of data, the concept was to take what was generally accepted as routine rates of delivery for intravenous bolus administration and, based upon experience, plot a way forward for volumes considered to be physiologically acceptable as the delivery period is extended. This experience is linked to the known fluid homeostasis scenario described in Chapter 1 and whole blood volumes (Table 1.2) and fluid intake and output (Table 1.4). From these data it was estimated that 10–20% of total blood volume could be safely administered by intravenous bolus injection of an isotonic solution and that this percentage volume would increase with time over a 24-hour period such that the total volume over 24 hours would approximate to an estimated average fluid output by the chosen species.

Table 3.3 Rates and Volumes of Intravenous Infusion Delivery – Non-Rodents

Duration of administration							
1–2 min	3–10 min	For each minute up to 1 hour[a]	Up to 1 hour[1]	For each hour up to 6 hours[b]	Up to 6 hours[1]	For each hour up to 24 hours[c]	Up to 24 hours or longer[1]
Maximum dose volume (mL/kg)[2]							
10	15	+0.1	20	+2	30	+1.5	60
Rate (mL/kg/min)				**Rate (mL/kg/hour)**			
5(10)	1.5(5)	-	0.33(2)	-	5	-	2.5

[1] Volumes specified indicate maximum (total) dose volume (mL/kg) for specified time period.
[2] Wherever possible the dose volume should be kept to the minimum practical.

To achieve this, the bolus rates were first set over a two-minute period, followed by a natural extension into the area of 'slow bolus injections' over a period of 3 to 10 minutes. Generally speaking, this period of intravenous dosing up to a duration of 10 minutes could be considered a 'bolus' administration and is often carried out in hand-restrained animals. Thereafter, the definition of when bolus intravenous administration ends and intravenous infusion administration begins becomes a matter of debate. In fact, the simplest of definitions/descriptions, which are often the less tainted, is that an intravenous **injection** is a delivery that takes a matter of seconds (less than a minute) and that any intravenous delivery taking longer than this becomes an **infusion**. This categorisation appears to hold well for all parenteral delivery systems such as vascular, subcutaneous, intramuscular, and intradermal. The definition also avoids the inclusion of unnecessary complications such as the use of surgically implanted catheters, the use of mechanical pumping devices, or the need for patient compliance.

Consequently, Tables 3.2 and 3.3 show start points for short vascular infusions (slow injections) of up to 10 minutes and, thereafter, slow predicted safe escalations for longer-term infusions of up to 24 hours. Apart from the first 10 minutes, there were two other defined 'milestone' time points during the first given 24-hour time period of an infusion delivery: +1 hour and +6 hours. Between these milestone time points, it was considered that escalations in infusion rates could be linear with time since the elimination process to compensate for hypervolaemic conditions would itself be directly dependent on the rate of increase in infused fluid volume. This is, of course, true up to the point of overload on the body fluid homeostasis process, which itself will be dependent on the properties of the infusate (discussed in detail in previous chapters), as well as maintenance of normal fluid elimination and control functions (mainly via the renal, gastro-intestinal, and respiratory systems). The details of the latter processes are outside the remit of this publication but are well-documented phenomena published in many excellent references.

The data in Tables 3.2 and 3.3 are represented graphically in Figure 3.1. The presentation of intravenous infusion rate data in this way demonstrates a number of features in what is the perceived physiologically acceptable pattern of delivery. The phase of delivery at **A** shows a lower start-point for intravenous bolus injections in smaller animals (rodents and rabbits) than for the non-rodent species. This is probably a consequence of anatomical features of the actual veins used for delivery in these species, which have different capacities for expansion to accommodate the introduction of rapid large volumes, for example tail veins for rodents and ear veins for rabbits compared to the more accommodating cephalic, saphenous, and jugular veins of the larger animals. At **Phase B** of the curve there is

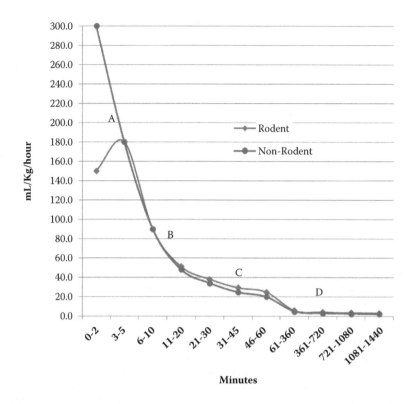

Figure 3.1 (See colour insert.) Rate of delivery versus infusion duration for rodents and non-rodents.

a rapid drop in the potential infusion rate over a period of up to 30 minutes in duration. However, during this period, the total infusion volume over the period is increasing slightly in line with the increasing capacity for the elimination of excess fluid over the time period. In **Phase C**, the intravenous infusion periods between 30 minutes and one hour, though the maximum infusion rate is considered to plateau, the total volume of infusate delivered increases significantly as the physiological capability over extended time periods for maintaining homeostasis becomes prevalent. **Phase D** of the curve relates to intravenous infusions of longer duration than one hour and, during the phase, it is the consideration of researchers that the physiologically acceptable rate of intravenous delivery falls further. This is because the total volume delivered increases as the time period of infusion increases. It is important, therefore, that the total volume of infusate within a 24-hour period be considered physiologically acceptable to a normovolaemic animal model (often related to normal daily fluid intake and elimination data) and that the subject is considered to have the capability to eliminate the excess fluid and maintain normal body fluid homeostasis within physiologically acceptable ranges.

Clearly these ranges are meant only as guidelines for intravenous infusion of xenobiotics and are subject to the influences of all the physico-chemical properties of infusates discussed previously. Also, these proposals are presented as average volumes and rates of delivery based on the infusion of a non-toxic, isotonic solution such as 0.9% sodium chloride. As such, it is expected that much higher rates of delivery and total daily volumes could be safely achieved should formulation constraints and toxicological requirements demand it.

References

Bivin SW, Crawford MP and Brewer NR. 1979. Morphophysiology. In: *The Laboratory Rat*, Baker HJ, Lindsey JR and Welsbroth SH, (eds), p. 90. San Diego, Calif.: Academic Press.

Bunag RD and Butterfield J. 1982. Tail-cuff blood pressure measurement without pre-heating in awake rats. *Hypertension* 4: 898–903.

Chalifoux A, Dallaire A, Blais D, Larivière N and Pelletier N. 1985. Evaluation of the arterial blood pressure of dogs by two non-invasive methods. *Can. J. Comp. Med.* 49(4): 419–423.

Claussen G and Tyssebotn I. 1973. Intrarenal distribution of glomerular filtration in conscious rats during isotonic saline infusion. *Acta. Physiol. Scand.* 89: 289–295.

Diehl K-H, Hull R, Morton D, Pfister R, Rabemampianina Y, Smith D, Vidal J-M and van de Vorstenbosch C. 2001. A good practice guide to the administration of substances and removal of blood, including routes and volumes. *Journal of Applied Toxicology* 21: 15–23.

Guo Y, Lin CX, Lou WY, Long D, Lao CY, Wen Z, Lai ECh, Wang XJ, Li LQ and Qing X. 2011. Haemodynamics and oxygen transport dynamics during hepatic resection at different central venous pressures in a pig model. *Hepatobiliary Pancreat. Dist. Int.* 10(5): 516–520.

Herrold EM, Goldweit RS, Carter JN, Zuccotti G and Bover JS. 1992. Non-invasive laser-based blood pressure measurement in rabbits. *Am. J. Hypertens.* 5(3): 197–202.

Hoefer IE, van Royen N, Buschmann IR, Piek JJ and Schaper W. 2001. Time course of arteriogenesis following femoral artery occlusion in the rabbit. *Cardiovasc. Res.* 49(3): 609–617.

Joint Working Group on Refinement. 2001. Administration of substances. *Laboratory Animals* 35: 19–41.

Krege JH, Hodgin JB, Hagaman JR and Smithies O. 1995. A non-invasive computerised tail-cuff system for measuring blood pressure in mice. *Hypertension* 25: 1111–1115.

Kumiko O and Yoshiaki Y. 1990. Noxious stimulation reduces blood pressure but not flow in the internal carotid artery as measured in rabbits. *Anaesth. Prog.* 37(1): 24–28.

Lukas VS. 1999. Volume guidelines for compound administration. In: *The Care and Feeding of an IACUC*, Podolsky ML and Lukas VS (eds), pp. 187–188. Boca Raton, Florida: CRC Press.

Mattson DL. 1998. Long-term measurement of arterial blood pressure in conscious mice. *Am. J. Physiol. Regul. Integr. Comp. Physiol.* 274: R564–R570.

Mignini MA, Piacentini E and Dubin A. 2006. Peripheral arterial blood pressure monitoring adequately tracks central arterial blood pressure in critically ill patients: An observational study. *Critical Care* 10(2): R43.

Morton D, Safron JA, Rice DW, Wilson DM and White RD. 1997. Effects of infusion rates in rats receiving repeated large volumes of saline solution intravenously. *Laboratory Animal Science* 47(6): 656–659.

Myrie SB, McKnight LL, van Vliet BN and Bertolo RF. 2006. Compensatory growth and blood pressure using non-invasive and telemetry techniques in Yucatan minipigs. *FASEB J.* 20: LB104.

Nelson NC, Drost WT, Lerche P and Bonagura JD. 2010. Non-invasive estimation of central venous pressure in anaesthetised dogs by measurement of hepatic venous blood flow velocity and abdominal venous diameter. *Vet. Radiol. Ultrasound* 51(3): 313–323.

Ochsner A, Colp R and Burch GE. 1951. Normal blood pressure in the superficial venous system of man at rest in the supine position. *Circulation* 3: 674–680.

Podolsky ML and Lukas VS, eds. 1999. *The Care and Feeding of an IACUC.* Boca Raton, Florida: CRC Press.

Scheepe JR, van den Hoek J, Jünemann KP and Alken P. 2007. A standardised minipig model for *in vivo* investigations of anticholinergic effects on bladder function and salivation. *Pharmacol. Res.* 55(5): 450–454.

Scheuermann-Freestone M, Freestone NS, Langenickel T, Höhnel K, Dietz R and Willenbrock R. 2001. A new model of congestive heart failure in the mouse due to chronic volume overload. *Eur. J. Heart Failure* 3: 535–543.

Tanaka K, Gotoh TM, Awazu C and Morita H. 2005. Regional difference of blood flow in anaesthetised rats during reduced gravity-induced by parabolic flight. *J. Appl. Physiol.* 99(6): 2144–2148.

Xu H, Aibiki M, Seki K, Ogura S and Ogli K. 1998. Effects of dexmedetomidine: An alpha2-adrenoceptor agonist on renal sympathetic nerve activity, blood pressure, heart rate and central venous pressure in urethane anaesthetised rabbits. *J. Auton. Nerv. Syst.* 71(1): 48–54.

chapter four

Formulation considerations

Co-author: Kevin Sooben
AstraZeneca, USA

Compounds intended for vascular administration are rarely administered solely as pure chemical substances but are almost always given in formulated preparations. These can vary from very simple solutions to complex drug delivery systems, and the use of appropriate additives or excipients in the formulations to provide varied and specialised pharmaceutical functions. It is the extra ingredients that, amongst other things, solubilise, suspend, thicken, preserve, and emulsify. The principal objective of the dosage form design is to achieve a predictable therapeutic response to a compound or active ingredient included in a formulation that is preserved in quality and consistency during large-scale manufacture. In this chapter, all the essential elements of the preparation of a formulation intended for vascular delivery are considered. This delivery is predominantly into venous blood via a peripheral vessel in an appendage.

Introduction

The ideal formulation for delivering compounds by infusion is a simple aqueous solution delivered at physiological pH and osmolality. In practice, formulations vary from simple solutions to complex drug delivery systems that use excipients performing various pharmaceutical functions. It is therefore vitally important that the excipients enable the administration of compounds without confounding their effects.

The principal objective of the dosage form design is to achieve a predictable response to a compound or active ingredient included in a formulation preserved in quality and consistency during large-scale manufacture. This chapter considers the essential elements of the development and preparation of a formulation intended for intravenous delivery. This is predominantly into venous blood via a peripheral vessel in an appendage. The formulation of such a product involves consideration of a number of interrelating factors covered in this chapter, including

- Formulation selection strategy
- Study design and species/strain
- Properties of the compound
- Formulation considerations
- Strategies for delivering poorly soluble compounds
- Unwanted formulation effects
- Pharmacokinetics and disposition
- Excipient toxicity
- Injection site reactions and haemolysis
- Strategies for dealing with poor stability
- Sterility

Formulation selection strategy

As compounds become more challenging to deliver for reasons such as increasing lipophilicity, poor solubility and stability, and poor local tolerability, there is a greater chance that an infusion formulation may fail for some of the reasons discussed in this chapter. However, these problems are frequently avoided by following basic principles that apply to all formulation development. These are

- **Preparation:** Understanding as much as possible the properties of the compounds, the objectives of the study, how the study will be conducted, and the species of interest
- **Keep it simple:** Using as simple a formulation and manufacturing process as possible, minimizing the number of steps involved, and using well-understood excipients within their normal dosing limits to reduce the number of opportunities for the formulation to fail
- **Have a plan:** Commencing a study with knowledge of the formulation risks associated with a compound and study to enable anticipation of potential problems and a mitigation plan to be put in place should issues occur

This chapter briefly describes these considerations and should be used as a guide with the emphasis on understanding their impact on intravenous delivery. These considerations are not exhaustive and do not constitute an in-depth instruction manual on formulation development. The reference sources used in this chapter contain detailed information, should more depth be required.

The characteristics of an ideal infusion formulation are

- Solution of compound well below (<80%) saturation solubility at all pHs encountered
- Physiological pH 7.4
- Low buffer capacity (<0.1)
- Iso-osmotic 280–290 mOsm/kg
- Non-toxic, non-irritant, and does not interfere with pharmacokinetics
- Non-precipitating upon infusion

Use of a formulation possessing these properties would negate the need for further formulation assessment, as they would render a formulation suitable for the majority of infusion studies. However, these properties are rarely all achieved at the same time. The consequences and strategies to deal with them are described in the remainder of this chapter.

It is important to understand that there are no standard vehicles, volumes, maximum excipient levels, etc., as they are not independent variables. Information such as the LD_{50}s found in the excipient datasheets, the typical excipient levels in Figure 4.1, and the guideline administration volumes in Table 4.1 (abstracted from Diehl et al. 2001) can be used to guide vehicle selection. It may be necessary to perform vehicle tolerability studies if the use of new excipients, extreme levels, or new combinations are planned, and it is advisable to attempt to minimise dose volumes and durations in these situations. Formulation development should be performed on a case-by-case basis according to the needs of the study.

Study design and species/strain

It is vital to understand the purpose of the formulation during its design phase. Key information that should be collected includes dose, volume, frequency, length of administration, and species/strain.

Formulation development may be straightforward for highly soluble and stable molecules that require little pharmaceutical intervention to ensure safe delivery. However, the formulation strategies and excipients used to overcome some challenges may inadvertently introduce bias or generate erroneous results if the type of study is not taken into account. There are three types of study that most pre-clinical infusion studies fall into: pharmacokinetic, pharmacodynamic, and toxicity studies (Li and Zhao 2007).

Pharmacokinetic (PK) studies are performed to investigate the absorption, distribution, metabolism, and excretion (ADME) of compounds that are administered. Pharmacodynamic studies examine the efficacy of physiological effects. Toxicology studies investigate a number of different parameters, including safety pharmacology, single/repeat dose toxicity, and carcinogenicity.

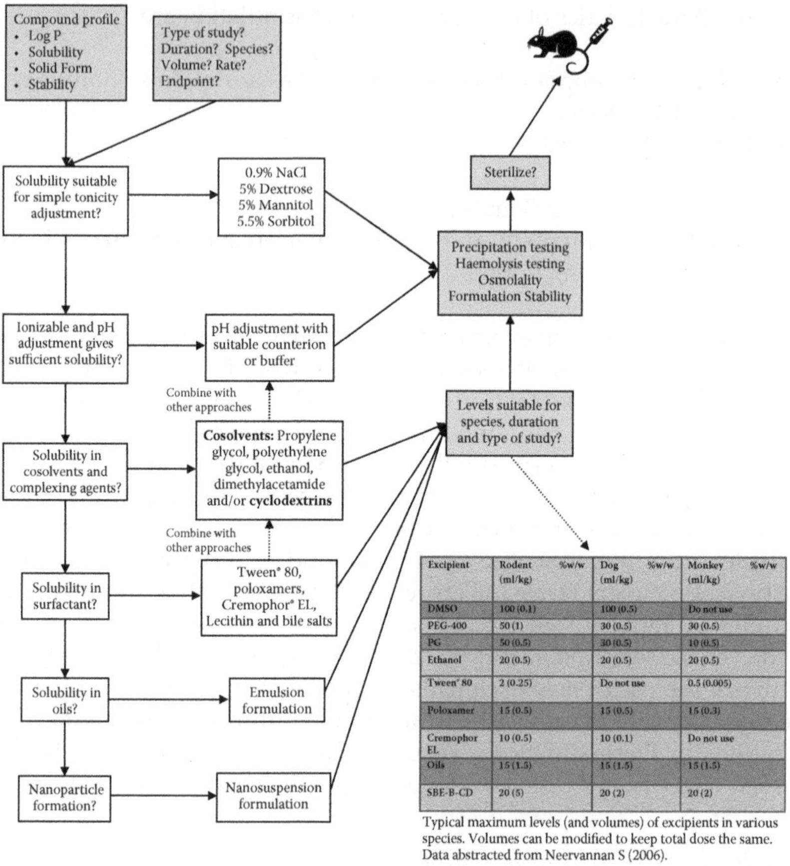
Compound profile
• Log P
• Solubility
• Solid Form
• Stability
Type of study? Duration? Species? Volume? Rate? Endpoint?
Solubility suitable for simple tonicity adjustment?
0.9% NaCl
5% Dextrose
5% Mannitol
5.5% Sorbitol
Sterilize?
Precipitation testing
Haemolysis testing
Osmolality
Formulation Stability
Ionizable and pH adjustment gives sufficient solubility?
pH adjustment with suitable counterion or buffer
Combine with other approaches
Levels suitable for species, duration and type of study?
Solubility in cosolvents and complexing agents?
Cosolvents: Propylene glycol, polyethylene glycol, ethanol, dimethylacetamide and/or cyclodextrins
Combine with other approaches
Solubility in surfactant?
Tween® 80, poloxamers, Cremophor® EL, Lecithin and bile salts
Solubility in oils?
Emulsion formulation
Nanoparticle formation?
Nanosuspension formulation
Excipient | Rodent (ml/kg) %w/w | Dog (ml/kg) %w/w | Monkey (ml/kg) %w/w
DMSO | 100 (0.1) | 100 (0.5) | Do not use
PEG-400 | 50 (1) | 30 (0.5) | 30 (0.5)
PG | 50 (0.5) | 30 (0.5) | 10 (0.5)
Ethanol | 20 (0.5) | 20 (0.5) | 20 (0.5)
Tween® 80 | 2 (0.25) | Do not use | 0.5 (0.005)
Poloxamer | 15 (0.5) | 15 (0.5) | 15 (0.3)
Cremophor EL | 10 (0.5) | 10 (0.1) | Do not use
Oils | 15 (1.5) | 15 (1.5) | 15 (1.5)
SBE-B-CD | 20 (5) | 20 (2) | 20 (2)
Typical maximum levels (and volumes) of excipients in various species. Volumes can be modified to keep total dose the same. Data abstracted from Neervannan S (2006).

Figure 4.1 The outline of a formulation selection strategy that also serves as a map of this chapter, summarising important considerations for developing a formulation.

Table 4.1 Administration Volumes Considered Good Practice (and Possible Maximal Dose Volumes)

Species	i.v. Bolus (mL kg^{-1})	i.v. Slow injection (mL kg^{-1})
Mouse	5	(25)
Rat	5	(20)
Rabbit	2	(10)
Dog	2.5	(5)
Macaque	2	Unknown
Marmoset	2.5	(10)
Minipig	2.5	(5)

It can be seen that these types of studies will have different dose requirements, endpoints, frequency of dosing, duration of dosing, and study lengths. It is, therefore, important to consider how these might affect the choice of formulation.

Dose

The dose is probably the single most important factor that affects how a compound may be formulated and administered. The solubility of the compound will decide the choice and level of solubility-enhancing excipients/strategies that may need to be employed. These strategies are described later in this chapter.

Endpoint, frequency, duration, and species

A thorough understanding of the endpoint of an infusion study is important, as it will affect the choice of excipients in a formulation. The purpose of PK studies is to assess the intrinsic ADME properties of compounds; it is therefore important that formulations do not interfere with the assessment of these properties. This is often also a consideration for pharmacology and toxicity studies. Excipient ADME effects are compound dependent and are related either to the way the excipients interact with the compound directly, or through the enhancement or inhibition of elimination and/or distribution processes.

Different species/strains will react differently to commonly used excipients. Excipients are subject to their own metabolism and excretion pathways, which may differ between species/strain, and some excipients may even cause toxic effects upon single-dose administration and/or multiple-dose administration. They may be subject to a maximum daily administrable dose that must be considered when considering frequency of administration.

It is important to select a formulation type that best matches the study to be conducted. Following are a few examples of the kind of formulation decisions that could be encountered in practice.

- A formulation with a high cosolvent level may be appropriate for single-dose PK study, but multiple dosing would cause toxicity that precludes its use in a toxicity study.
- Conversely, a formulation that has an impact on the PK parameters for a compound may be suitable for a toxicity assessment where the objective is maximum exposure, but may not be suitable for PK studies.
- A formulation vehicle that causes a mild and well-understood toxicity that is unrelated to the mechanism of action of a compound may be suitable in studies with a well-defined pharmacology endpoint.

- Early PK studies may require the use of 'generic' formulations for high-throughput testing of compounds. In these cases, it may be appropriate to use excipients with known toxic/tolerability and PK effects to study gross differences between compounds, whilst ensuring that the majority of compounds are adequately solubilised.

The pharmacokinetic and toxic effects referred to above are discussed in further detail later in this chapter.

Properties of the compound

The study of the physico-chemical properties of a compound is known as *preformulation*. Preformulation studies facilitate the rational selection of an appropriate formulation for its delivery via an infusion.

There are many aspects of a compound that can be studied during preformulation. The most important from an infusion perspective are a compound's solid form, partition coefficient (log P), solubility and pKa, and stability. A brief description of these preformulation studies and their relevance to infusion formulation is given here. Readers are referred to the many reference works, such as Gibson (2001), on the subject of pharmaceutical preformulation for further details and the methods used to generate this information.

Solid form

Compound solids can exist in a number of different states. Amorphous solids have randomly arranged molecules. Conversely, crystalline solids are packed in an ordered state. Furthermore, a crystalline solid can exist in different ordered arrangements referred to as *polymorphs*, as shown in Figure 4.2.

Amorphous materials exist in a higher energy state compared to their crystalline forms. Polymorphs also exist in different energy states. The polymorph with the lowest energy state is referred to as the stable polymorph, and it always has the lowest solubility of all forms under the same conditions.

Characterisation of the solid form is critical for formulation because it can have a huge impact on its solubility and physical stability. Compounds that are formulated from amorphous or metastable polymorphs can have a tendency to convert to the stable polymorph over time. The implication is that precipitation will occur if the concentration in the formulation is above the solubility limit of the stable polymorph.

It is desirable to use a stable polymorph to ensure predictable physical stability in most instances. However, there are circumstances where it is acceptable to use amorphous or metastable forms, such as biological

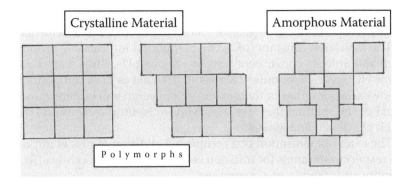

Figure 4.2 Crystalline and amorphous forms.

molecules or when the material behaviour is well understood. Understating the solid-state properties of the compound thus allows a better idea of the formulation risks especially with respect to physical stability, that is, precipitation on storage.

A number of analytical techniques such as optical microscopy (visual confirmation of crystallinity), differential scanning calorimetry (melting point), and X-ray powder diffraction (structural differences between forms) are used to investigate solid form. These are beyond the scope of this chapter, and readers are referred to sources such as Newman and Byrn (2003) for further information.

Lipophilicity

The lipophilicity (logP) of a compound indicates its hydrophilic (water-loving) or lipophilic (fat-loving) characteristics. It is the ratio of the concentration of a compound at equilibrium between an organic phase (usually a solvent such as octanol) and an aqueous phase. It is log scale where higher numbers correspond to more lipophilic compounds. A logP of zero would correspond to equal octanol and aqueous solubility, and each unit increase represents a tenfold relative increase in octanol solubility. The logP can be a useful early indicator of potential aqueous solubility issues (Lipinski 2000) and helpful in deciding on formulation strategy for poorly water-soluble compounds.

Solubility and pKa

The single most important factor in determining the success of an aqueous solution formulation is, arguably, its intrinsic solubility. The intrinsic solubility of a compound is measured with the stable polymorph at a pH where it is un-ionised. Its determination along with a measure of the ionisation constant (for basic or acidic compounds) can give an early indication of potential strategies for increasing its solubility in a formation.

The ionisation constant (pKa) of a compound indicates its propensity to separate into its constituent ions at various pHs. Thus a strong acid, such as HCl (pKa –7), is ionised across all pHs, but as most compounds are either weak acids or bases, their extent of ionisation will be determined by the pH of the solution they are dissolved in. Neutral compounds do not exhibit pH-dependent ionisation.

The extent of ionisation of a compound at different pHs is important with respect to solutions for infusion because it can significantly affect the solubility and stability of a compound.

There are a number of methods for obtaining pH solubility data. One method is to measure the saturation solubility in a system of buffers across a pH range. There are various techniques to obtain this data, as reported in Higuchi and Connors (1965) and Kumar (1997). Another method is to measure single point solubility (usually the intrinsic solubility) and predict the solubility-based theoretical considerations.

Figure 4.3 shows an overlay of a predicted solubility profile with actual measured values for a weakly basic compound. This example shows good correspondence between predicted and measured solubility values, and it can be seen that this data could be used to help select the optimal formulation pH to obtain the desired solubility. A weakly acidic compound would display a mirror image in its pH solubility profile.

It must also be noted that acids and bases will increase in solubility exponentially until a maximum is achieved, after which further increases or decreases in pH do not increase the solubility for acids and bases, respectively. Additionally, the type and concentration of buffers used can influence the measured solubility, leading to potentially misleading results that lead to poor-quality formulations. The type of buffer used may lead to the formation of more or less soluble salts depending on how the counterions used in the buffer interact with the compound of interest. This effect is observed to a certain extent in Figure 4.3 with the difference in solubilities between the lactic acid and sulphuric acid at the same pH. Any salt could also be susceptible to the common ion effect, where increases in concentration of the ion used to form the salt causes a decrease in solubility of the compound.

Readers are referred to Gibson (2001) for further theoretical depth on solubility and pH.

Salts

If a compound can be ionised, then it has the potential to form a salt through dissociation and an ionic interaction with an oppositely charged counterion. There are many reasons to form a salt of a compound, which are discussed briefly in this section.

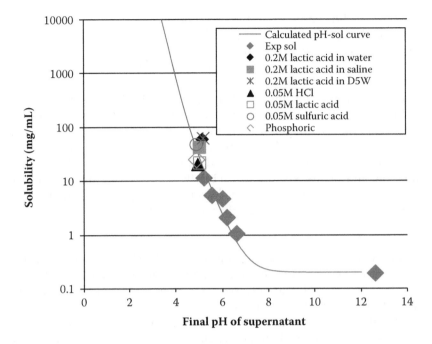

Figure 4.3 (See colour insert.) Predicted and measured solubilities for a basic compound.

The counterion used to form a salt can have an impact on its solubility, but not in every case. A salt may have little or no impact on the solubility, but salts are still utilised to enhance formulation by facilitating manufacture through an enhancement in dissolution rate. This is caused by the creation of a high micro-climate pH for basic counterions (and low for acidic) when solid particles of a salt of a compound begin to dissolve. This creates a transient region of high compound solubility and facilitates dissolution.

Salts are also used to solve potential chemical stability issues caused by hygroscopic compounds that pick up moisture from the atmosphere in their free state, and to improve the solid-state properties of compounds, such as allowing the formation of a stable polymorph where only amorphous free form can be isolated.

Gould (1986) and Gibson (2001) discuss the use and formation of salts in detail.

Stability
The stability of a compound in solution and solid states can be important for a number of reasons. It can dictate

- how a solid compound is handled and stored
- how the compound is formulated
- how the compound is administered
- the shelf life of the compound and formulation

In addition, it may be important to understand what the degradation products are, as they may have a pharmacological or toxicological impact.

The solution stability is usually the immediate concern from an infusion perspective, though the chemical stability in the solid state can also be important. The most important degradation routes are hydrolysis, oxidation, and photolysis. These are briefly described in this section, but the mechanisms of degradation are beyond the scope of this chapter. See Connors et al. (1986) for a further discussion of this area.

Hydrolysis

Compounds that degrade by hydrolysis do so by the addition of nucleophilic molecules such as water or hydroxyl ions (OH⁻) and can be affected by factors such as pH (concentration of H^+ and OH⁻), ionic strength, the type of buffer, and other additives.

The pH is usually a key factor that determines the degradation rate of a compound, and it is therefore useful to have an understanding of the pH/hydrolytic stability relationship of a compound before the final pH of the formulation is selected.

Oxidation

Many compounds degrade by oxidation. Unlike hydrolysis, this is not usually directly affected by pH but occurs in the presence of oxygen. The rate of reaction can be affected by light, temperature, and the presence of impurities that promote oxidation through the formation of free radicals.

Oxidation studies are usually performed by the addition of hydrogen peroxide to the solution of the compound of interest.

Photochemical degradation

Most compounds absorb light of different wavelengths to varying degrees. This is their absorption spectra, which is an indication that the molecular bonds absorb light energy. This can lead to breakage of these bonds and to photochemical degradation.

Most compounds with photostability issues exhibit significant absorption above a wavelength of 330 nm (Albini and Fasani 1998), but not all compounds that absorb above this wavelength will degrade.

There are a number of chemical groups that absorb light and can be susceptible to degradation, and an examination of the chemical structure of the molecule should be the first step in assessing the degradation risk. Refer to Albini and Fasani (1998) for a detailed description.

During preformulation studies it is important to perform the necessary experiments to enable the photochemical degradation risk of both the solid form and solutions so that appropriate measures can be taken to protect against it.

Strategies for dealing with poor solubility

The two generally accepted main causes of poor solubility are high lipophilicity and high crystalline lattice energies, the so-called 'grease-ball' and 'brick-dust' type compounds. In general there is an inverse relationship between lipophilicity and the solubility of a compound's un-ionised form, so the most lipophilic molecules tend to be the least soluble (Box and Comer 2008). However, some 'brick-dust' poorly water-soluble compounds are not particularly lipophilic. Instead, they exhibit a high-energy barrier to liberating molecules from their crystals for dissolution.

The preformulation studies described earlier in this chapter give an indication of the reason for poor solubility and its potential mitigation.

pH Adjustment

If a compound has an ionisable centre, then pH adjustment is often the most simple and effective method to achieve a solution. This can be accomplished by simple addition of acids or bases to a formulation. This leads to an increase in solubility through an increase in ionisation and sometimes the formation of 'in-situ' salt with the counterion of the acid or base used. Figure 4.3 shows a typical pH-solubility profile for a weak base. The selection of the optimal pH will be based on physiological considerations for biocompatibility and the solubility/stability studies performed during preformulation.

Formulations should be dosed as close to physiological pH as possible, but in practice pHs in the range 4–8 are commonly administered. Formulations at greater extremes (pH 2–11) can be dosed (Sweetana and Akers 1996), but these systems must be well understood in order to avoid potential adverse effects (see the Unwanted Formulation Effects section).

Buffers

Buffers are used as a means to maintain a solution at a target pH and resist pH changes when acids or bases are added to the solution. Most buffer solutions consist of a weak acid and its conjugate base (salt of the weak acid). Buffers should be used where control of pH is required to obtain optimal solubility and/or stability. Table 4.2, reproduced from Gibson (2001), gives examples of buffers used in parenteral products along with the buffering pH range at which they work.

Table 4.2 Buffers Used in Parenteral Products

Buffer	pH Range
Acetate	3.8–5.8
Ammonium	8.25–10.25
Ascorbate	3.0–5.0
Benzoate	6.0–7.0
Bicarbonate	4.0–11.0
Citrate	2.1–6.2
Diethanolamine	8.0–10.0
Glycine	8.8–10.8
Lactate	2.1–4.1
Phosphate	3.0–8.0
Succinate	3.2–6.6
Tartrate	2.0–5.3
Tromethamine (TRIS, THAM)	7.1–9.1

Acids and bases produce H^+ and OH^- ions, respectively, decreasing or increasing the pH of water. Buffer systems resist large pH changes because the weak acid and its conjugate base neutralise added acids or bases. The buffer capacity is the number of equivalents of acid or base required to change the pH of 1 L of solution, and it is controlled by concentration of the buffer system components (doubling the total molar concentration will double the buffer capacity at a given pH) and the ratio of the acid and its conjugate base. Buffer capacity is maximal when the pH = pKa of the acid, that is, when there is an equal concentration of the acid and its conjugate base.

A higher buffer capacity may be important in maintaining the compound solubility at a certain pH and reduce precipitation, but this in turn can affect the tolerability of a formulation (see the Unwanted Formulation Effects section). In general, it is advisable to use a buffer as close to physiological pH with a minimal buffer capacity to reduce the risk of buffering the animal and changing its physiological pH upon intravenous administration. Readers are directed to Kaus (1998) for further theory and practical aspects of the use of buffers.

Cosolvents

Cosolvents such as ethanol, propylene glycol, glycerin, dimethyl sulfoxide, and polyethylene glycol (Sweetana and Akers (1996), and the excipient data-sheets in the Annex) are used, frequently in combination with pH adjust-ment, to increase the solubility of poorly soluble compounds for intravenous delivery. They often have a high solubilising capacity and are easy to prepare.

Compounds dissolved in cosolvents/aqueous mixtures exhibit a logarithmic increase in solubility as the concentration of cosolvent increases, as shown in Figure 4.4 (Yalkowsky 1999). One of the implications of this is the risk of precipitation upon *in vivo* dilution (see Unwanted Formulation Effects section).

The solubility in a particular cosolvent is dependent upon the polarity (and lipophilicity) of the compound with respect to the cosolvent and the diluents used (usually water). The effect of this can be seen in Figure 4.5 where two different compounds exhibit different solubility profiles in decreasing fractions of ethanol and PEG400.

There are number of theoretical and practical methods for investigating the solubilisation potential for compounds in cosolvents, which are described in Sweetana and Akers (1996).

Complexing agents

Lipophilic compounds are capable of forming an association with large (>1000 MW) water-soluble molecules with a hydrophobic core. This association is called *complexation* and can be utilised to increase the apparent solubility of insoluble compounds in a formulation. Cyclodextrins are such molecules and are used to deliver poorly soluble molecules by the intravenous route.

Cyclodextrins are cyclic oligosaccharides composed of glucopyranose molecules and are manufactured from the enzymatic conversion of starch. They have a hydrophobic core cavity with a hydrophilic outer surface. There are a number of different types of cyclodextrin that differ in their number of glucopyranose units and substituents. Owing to their potential for renal toxicity (see the Unwanted Formulation Effects section), two water-soluble derivatives have been developed. Sulfobutyl ether and hydroxylpropyl derivatives of β-cyclodextrin, sulfobutylether-β-cyclodextrin

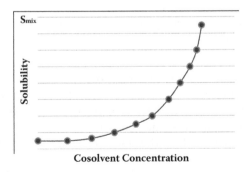

Figure 4.4 Solubility dependence on cosolvent concentration.

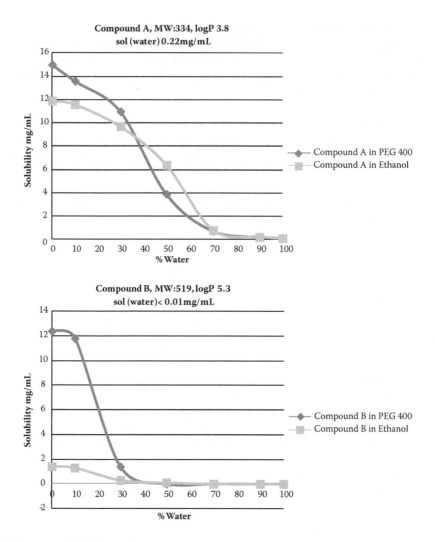

Figure 4.5 Solubility of two compounds in decreasing concentration of ethanol and PEG 400.

(SBE-β-CD, Captisol®) and hydroxypropyl-β-cyclodextrin (HP-β-CD) are used increasingly in pre-clinical and clinical parenteral formulations because of their low toxicity. It is for these reasons that these two cyclodextrin derivatives are the focus of this chapter rather than any other derivatives. See Stella and Rajewski (1997) for a review of the other derivatives.

High concentrations of cyclodextrin are usually required to solubilise a compound through a 1:1 association, as shown in Figure 4.6

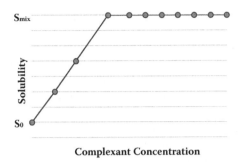

Complexant Concentration

Figure 4.6 Solubility dependence on complexant concentration.

(Yalkowski 1999). However, higher-order complexes have been noted and are required for some compounds (Shi et al. 2009). Care must be taken to choose the correct proportion of cyclodextrin and to understand the binding to compounds to avoid inadvertently altering the PK properties of compounds. This concept is covered in the Unwanted Formulation Effects section.

The general range of concentrations that are used in preclinical infusion studies is 10–40% w/v. To minimise the amount of cyclodextrin in a formulation, it is often combined with the other techniques described in this section in order obtain an additive or synergistic effect.

There are no exact methods to predict the level of solubility enhancement from complexation with cyclodextrins, but generally a monosubstituted phenyl ring or other non-polar group of similar size and shape is best located a few atoms away from an easily protonated basic group (Shi et al. 2009). Differences between the capacity of SBE-β-CD and HP-β-CD may be due to the size of the hydrophobic cavity and the negative charge present on SBE-β-CD at physiological pH, which can adversely affect its ability to bind with negatively charged compounds.

Unlike many of the excipients described in this section, both HP-β-CD and SBE-β-CD are subject to licensing restrictions on their clinical use from Janssen/National Institutes of Health and CyDex Pharmaceutical, Inc. (Lenexa, KS). As a license is needed in order to commercialise any product using either of these cyclodextrins, thought must be given to their pre-clinical use as part of a broader formulation strategy.

Surfactants

Surfactants can be useful tools for enabling the intravenous formulation of poorly water-soluble compounds. They are also used to decrease the precipitation potential and increase chemical stability, but can cause other toxic effects and affect PK parameters, as discussed in further detail elsewhere in this chapter.

Surfactants derive their beneficial effects from being amphiphilic (dual nature), having both hydrophilic (water-loving) and lipophilic portions within the same molecule. This leads to the formation of self-assembled structures in aqueous media called *micelles* once a minimal surfactant concentration is achieved. When this critical micellar concentration (CMC) is reached, compounds with the right physico-chemical properties (generally lipophilic) can enter and concentrate inside the micelles, as shown in Figure 4.7.

When a surfactant is effective it generally exhibits a linear increase in the solubilisation capacity with increasing concentration once the CMC has been reached, as shown in Figure 4.8 (Yalkowsky 1999).

If a compound cannot be formulated using pH adjustments, cosolvents, complexation, or a combinations of these, the compound is considered 'challenging' (Strickley 2004). This is where surfactants are sometimes used to solubilise some of the most water-insoluble compounds. The number of surfactants that are used for intravenous infusion formulations is limited, and they are all non-ionic. These include Cremophor® EL, Cremophor RH60, lecithins, poloxamers, and polysorbate (Tween®) 80. Ionic surfactants (e.g. sodium dodecyl sulfate) are generally more toxic owing to their impact on biological membranes.

There are no officially published upper concentration limits for the use of surfactant in pre-clinical infusions, as there are number of interdependent factors, including LD_{50} and tolerable concentrations, that determine what can be used. Sweetana and Akers (1996) and Strickley (1999) describe clinical examples of the use of these surfactants. The excipient datasheets in the Annex and the Unwanted Formulation Effects section also contain further details on LD_{50} and toxicities that are encountered with surfactants.

Mixed micelles

Administration of poorly soluble compounds in mixed micelles represents a variation on the use of surfactants. They are usually composed of a phospholipid and a bile salt. Bile salts are produced endogenously and reported to cause haemolysis and irate veins when administered alone. However, they are reported to be better tolerated than on their own, and also better tolerated than the synthetic surfactants described above, when delivered in a mixed micellar system (Bittner and Mountfield 2002).

The solubilisation capacity depends on a number of parameters, including the compound of interest, the phospholipid uses, and the ionic strength (Sweetana and Akers 1996).

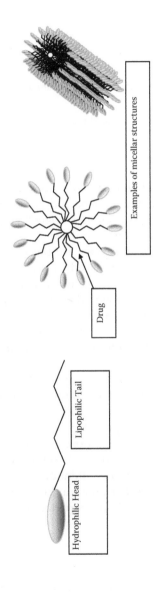

Figure 4.7 (See colour insert.) Schematic structure of surfactant molecules and types of micelle that may be formed.

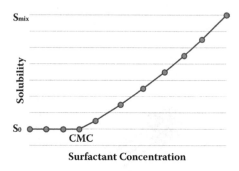

Figure 4.8 Solubility dependence on surfactant concentration.

Emulsions systems

Parenteral emulsion formulation systems are oil-in-water with lipid as the internal droplet phase and a dispersed aqueous phase (Figure 4.9). Typically, up to 30% of a triglyceride-rich vegetable oil such as soybean is used as the basis for an emulsion formulation.

Mixtures of oil droplets and water are thermodynamically unstable and require emulsifying agent(s) that reduce interfacial tension to prevent coalescence of the drops. These are usually lecithins, but non-ionic surfactants are sometimes also used. Mechanical energy is also required to produce sufficiently small droplets. Droplets should be less then 1000 nm to reduce the risk of emboli, but it is also recommended that the droplet size be less than 200 nm to reduce uptake in the liver and spleen (Bittner and Mountfield 2002).

A compound has to be sufficiently lipid soluble to be considered for formulation in an emulsion system. Emulsion formulations are more difficult and time consuming to develop, manufacture, scale-up, and sterilise effectively than the approaches discussed thus far but can be an effective way of delivering high doses of poorly water-soluble lipophilic compounds. Care must be taken when formulating an emulsion for intravenous delivery because of the potential of altering bio-distribution, as discussed in the Unwanted Formulation Effects section.

Refer to Hansrani et al. (1983) for more detail on the development and manufacture of IV emulsion formulations.

Nanosuspensions

There will be circumstances where the other approaches described in this chapter do not enable intravenous delivery of poorly soluble compounds. In these situations, it is possible to embrace poor solubility and deliver solids intravenously using nanosuspensions.

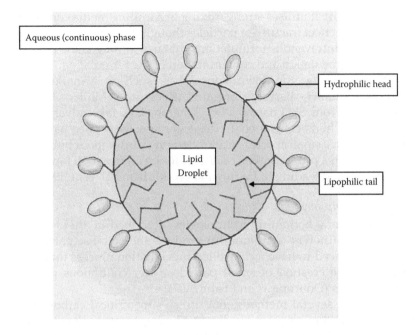

Figure 4.9 (See colour insert.) Schematic of an emulsion droplet.

Nanosuspensions are a formulation system where compound is presented as solid nanoparticles (smaller than 1 μm in diameter) dispersed in aqueous media with lipophilic or polymeric stabilisers. Particles for intravenous delivery need to be smaller than this to reduce the risk of emboli upon administration.

Nanosupensions can have a number of benefits over other formulation approaches. Since solubility in the formulation is no longer a concern, high compound loading (up to 40% w/v) may be achievable, as reported by Merisko-Liversidge et al. (2003). Unstable compounds may also benefit from being presented in a solid state rather than in solution where they may be susceptible to degradation by oxidation, hydrolysis, or photo degradation (Rabinow 2004).

There are a number of methods for manufacturing nanosuspensions that fall broadly into two categories: top-down and bottom-up. Top-down methods comminute larger starting particles to end up with smaller particles, and bottom-up methods build up particles by controlled precipitation.

Top-down methods

The key top-down approach (NanoCrystal®, Merisko-Liversidge et al. 2003) utilises milling technology to produce nanoparticles. The technology was developed by Alkermes (formerly Elan Drug Technologies,

Dublin, Ireland). It utilises strong mixing in a milling media consisting of hard beads to effect fracture of particles though impaction. One potential challenge for intravenous formulations that use milling is the possibility of media fragmentation and contamination of the formulation.

Other common top-down approaches, such as NanoPure® from Pharmasol GmbH, Berlin, Germany (Keck and Müller 2006) and Dissocubes® from SkyPharma PLC, London, UK, take advantage of a high-pressure homogenisation. In this technique suspended particles are forced through a small aperture at elevated pressure. Vapour bubbles form and collapse in a process called *cavitation* for creating a high force around the particles and leading to fragmentation (Müller et al. et al. 1998).

Bottom-up methods

There are several bottom-up technologies, described in Shi et al. (2009). They all use different techniques to produce and stabilise particles, but all share the need to have a controlled precipitation process that favours nucleation (the creation of small particles) over continuous growth of larger particles (Douroumis and Fahr 2007).

There are several methods that utilise supercritical carbon dioxide $(S-CO_2)$, where the compound is dissolved in an organic solvent or an oil-in-water emulsion (Bristow et al. 2001; Shekunov et al. 2006) and rapidly added to the $S-CO_2$, which acts as an antisolvent causing precipitation of compound nanoparticles. These methods give a dry product that requires reconstitution.

Another technology uses albumin-encapsulated nanoparticles and is the basis of the Abraxane® (Abraxis Bioscience, LLC, Los Angeles, CA, USA) paclitaxel nanosuspension formulation. In this system, high-pressure homogenisation is used to form an oil-in-water emulsion of compound dissolved in organic solvent in an aqueous phase containing albumin. The solvent is removed to form solid nanoparticles (Shi et al. 2009).

Two other technologies, NanoEdge (Baxter International, Deerfield, IL, USA) and Nanomorph (Abbott GmbH & Co., KG, Ludwigshaven, Germany), precipitate compound in the presence of lipid of polymer stabilisers. The former use a two-step process of precipitation from an organic solvent followed by high-pressure homogenisation. The latter generates amorphous nanoparticles.

Formulation considerations for nanoparticles

It is relatively easy to create nanoparticles. The main challenge is to stabilise them in a formulation that can be stored and administered without settling, aggregation, solid form transformations, or changes in the particle size distribution.

The inhibition of aggregation is usually achieved through the use of a limited range of parenterally acceptable stabilisers such as poloxamers, polyvinylpyrrolidones, phospholipids, and lecithin derivatives (Shi et al. 2009) that coat the particles and prevent close association. Unfortunately these stabilisers are some of the same excipients described earlier in this section that are used to increase the solubility of poorly soluble compounds. This can accelerate an effect known as *Ostwald ripening*, whereby smaller particles dissolve at the expense of larger particles, resulting in a gradual increase in the size of the particles over time.

It is important to achieve the right compound-to-stabiliser ratio. Merisko-Liversidge et al. (2003) suggested it should be in the 2:1 to 20:1 w/w range for compound:stabiliser.

The next step for nanosuspensions is sterilisation. This complexity of their production can make this challenging (see Sterility section).

Once these, often considerable, hurdles have been overcome, then a new set of potential hurdles may present themselves in trying to achieve the desired exposure profile. It is difficult to study the fate of nanoparticles once they are administered intravenously.

It has been reported that the solubility and the dissolution rate of the nanoparticles in the blood are key determinants of the distribution of compound once it has been administered. The large surface area of nanosuspension should maximise the dissolution rate and give similar PK to a solution, and this is reported to be the case for an itraconazole nanosuspension (~300 nm) compared to a HP-β-CD formulation (Mouton et al. 2006). However, when larger itraconazole nanoparticles (~580 nm) are administered they show a reduced Cmax and prolonged plasma half-life relative to a HP-β-CD formulation (Rabinow et al. 2007). The particles were taken up in the reticuloendothelial system in the lung, liver, and spleen, where they act as a depot, releasing the compound over time.

Nanosuspension may become a more important delivery vehicle as compounds follow the trend of becoming more poorly soluble. It is imperative to understand a compound and its fate upon administration before *in vivo* studies are started.

Unwanted formulation effects

Pharmaceutical excipients are often thought of as inert ingredients. In reality, it has been established since the 1960s (Brink and Stein 1967) that certain excipients can have a significant impact on the pharmacokinetic properties of a compound.

The strategies used to deliver increasingly poorly soluble compounds necessitates the use of formulation conditions and excipients that have been shown to have effects on the pharmacokinetics of the compounds and to reduce, potentiate, or directly cause unwanted local and systemic reactions.

Pharmacokinetics

The aim of this section is to indicate the potential impact that formulation excipients can have on the PK behaviour of compounds. Readers are referred to comprehensive articles by Bittner and Mountfield (2002) and Buggins et al. (2007), which cover this topic in great depth.

Cosolvents

There are many cosolvents that are suitable for parenteral use (see the excipient datasheets in the Annex for details). Most cosolvents affect the pharmacokinetic parameters of at least one compound in at least one species. These effects are analogous to drug–drug interactions (Bittner and Mountfield 2002).

Some of the more important effects for a range of cosolvents are shown in Table 4.3 and represent a small fraction of the literature on this subject. This table serves as a guide to the *potential* for these types of interactions and, as with drug–drug interactions, will be compound dependent. There are a number of articles that report no effect on some of the compounds not reported here.

Cyclodextrins

Cyclodextrins are often the subject of much debate, confusion, and concern as to their effects on the disposition of intravenously administered compounds. This stems from their mechanism of solubilisation and the belief that cyclodextrin-compound complexes remain intact after intravenous administration. The potential for this occurrence should always be considered, but the likelihood of encountering this issue in practice is low. There are at least 10 published studies (all reviewed in Buggins et al. 2007) that specifically address the effect of intravenously administered HP-β-CD or SBE-β-CD on the preclinical pharmacokinetics of a range of compounds. Only two report any effects on disposition, namely Frijlink et al. (2001) and Perry et al. (2006).

Frijlink et al. (2001) compared the disposition of two highly protein-bound (>99%) compounds in rats. Naproxen and flurbiprofen were each dosed in two intravenous formulations, one with and one without HP-β-CD. Compound levels in brain, liver, kidney, spleen, muscle, and plasma were measured at 10 and 60 minutes after dosing. Statistically significant higher concentration in brain, liver, kidney, and spleen were noted after 10 minutes with the HP-β-CD flurbiprofen formulation. However, they were no longer different after 60 minutes in all tissues except for slightly elevated levels in the brain.

This behaviour was explained by the difference in binding to HP-β-CD. The binding constant, as a measure of binding strength, was found to be unusually high for flurbiprofen compared to naproxen

Table 4.3 Effects of Cosolvents on Pharmacokinetics and Disposition

Cosolvent	Effect	Species or system	Drug	References
DMSO	Increased cerebral bloodflow	Dog	N/A	Kassell et al. 1983
DMSO	Increased heart concentration	Rat	Gentamycin	Rubinstein and Lev-El 1980
DMSO	Inhibited CYP1A2, 2C8, 2C9, 2D6, 2E1 and 3A4	Human liver microsomes	N/A	Chauret et al. 1998
Ethanol	Inhibited metabolism	Rabbit lung microsomes, rat	Benzo(a)-pyrene, propoxyphene	Kontir et al. 1986; Oguma and Levy 1981
Ethanol	Reduced clearance, increased clearance	Dog, pig	Cocaine	Parker and Laizure 2010; Kambam et al. 1994
Propylene glycol	Inhibited metabolism	Mouse liver fractions, rabbit liver microsomes	Acetaminophen, Dramamine	Snawder et al. 1993; Walters et al. 1993
PEGs	Reduced urine volume and increased creatinine elimination, protein in urine	Rat	N/A	Pestel et al. 2006
Glycofurol	Inhibition of hepatic metabolism, decreased elimination of zoxazolamine and hexobarbital followed by induced hepatic metabolism on repeat dosing	Dog, rat	Carbamazapine, zoxazolamine, hexobarbital	Loscher et al. 1995; Yasaka et al. 1978

(12500 M^{-1} and 2200 M^{-1} respectively). The higher-binding flurbiprofen was shown to statistically significantly decrease the level of protein binding when mixed with plasma *in vitro*. It was postulated by Frijlink et al. (2001) that this increased the amount of flurbiprofen available to the tissues.

Perry et al. (2006) found that a series of anti-malarials with unusually high binding constants to SBE-β-CD of greater than 10^5 M^{-1} had altered plasma PK (but not whole blood PK, owing to the association with red blood cells).

There have been many publications describing the practical significance of these findings, and all suggest that the effect cyclodextrins have on pharmacokinetics will not be significant, for the following reasons:

- HP-β-CD and SBE-β-CD are rapidly cleared, and therefore any potential effect from cyclodextrins could only be transient, as demonstrated in Frijlink et al. (2001).
- It has been shown that the binding constant must be about 10^5 M^{-1} or greater to have an impact on a compound's pharmacokinetics, as dilution in plasma, competitive binding to plasma protein, binding of endogenous agents to the cyclodextrin cavity, and uptake by tissues inaccessible to cyclodextrin will drive the majority of compound to the unbound state (Stella and He 2008; Stella and Rakewski 1997).
- Most compounds have binding constants between 10 and 2000 M^{-1}, with those above 5000 M^{-1} rarely observed (Kurkov et al. 2010).

Surfactants

Surfactants play an increasing role in parenteral formulations and have qualities that make them ideally suited to parenteral formulations, as described in other sections of this chapter. However, of all the excipients described in this section on pharmacokinetics, they have the greatest potential to have unwanted PK effects, and there are many publications that report the alteration of compound pharmacokinetics with the use of surfactants.

Table 4.4 summarises a small sample of the effects that have been reported in publications such as Buggins et al. (2007) and Bittner and Mountfield (2002) for the surfactants Cremophor EL, Tween (polysorbate 80) and Solutol® HS15. In contrast to the cosolvents, there is a smaller proportion of studies where no effects on the disposition of compounds is noted; but, as with the cosolvents, the effects appear to be compound dependent. It is advisable that careful consideration be given to understanding the potential effects of surfactants on the pharmacokinetics of a compound when planning intravenous infusion studies.

Table 4.4 Effects of Surfactants on Pharmacokinetics and Disposition

Surfactant	Effect	Species or system	Drug	References
Cremophor EL	Increased AUC, reduced Cl, VSS	Rat, baboon	BMS-310705, cyclosporin	Kamath et al. 2005; Jin et al. 2005; Kurlansky et al. 1986
Cremophor EL	Reduced Cmax and AUC, altered protein binding	Rat, human, *in vitro*	Paclitaxel	Bardelmeijer et al. 2002; Malingre et al. 2001; Sykes et al. 1994
Tween 80	Reduced Cl (through inhibition of P-glycoprotein)	Rat, rat isolated perfused liver	Digoxin, etoposide	Zhang et al. 2003; Ellis et al. 1996
Tween 80	Reduced then increased urinary excretion, increased faecal excretion, increased brain uptake	Mouse	Methotrexate	Azmin et al. 1985
Tween 80	Initial reduction in blood:plasma ratio, increased f_u after infusion	Rat, human	Docetaxel	Yokogawa et al. 2004; Loos et al. 2003
Solutol HS15	Alteration of PK with possible formation of ternary complex of drug, vehicle and plasma components, reduced electrophoretic mobility of HDL and LDL	Mouse, *in vitro*	C8CK	Woodburn et al. 1995

Emulsion systems

Emulsion systems offer similar advantages to surfactants in preventing precipitation of lipophilic compounds on intravenous administration (see section on precipitation). However, there are cases where lipidic emulsion formulations have had a significant impact on the pharmacokinetics of co-administered compounds (Bittner and Mountfield 2002).

The fate of compounds formulated in emulsions is closely linked to the fate of emulsion droplets. Therefore the circulation time (Takino et al. 1994) and the partition behaviour of a compound between oil droplets and plasma (Sakaeda and Hirano 1995) can significantly impact pharmacokinetics.

It has been shown that emulsion droplet size can affect its disposition, with a size less than 100 nm being optimal for enhanced circulation time. In much the same way as nanosuspensions, emulsion droplets are taken up into the reticuloendothelial system (Singh and Ravin 1986) in a size-dependent manner, with larger sizes more susceptible to uptake. If the compound does not partition out of the emulsion droplet rapidly enough, then it will be eliminated intact with the droplet. Sakaeda and Hirano (1995) suggest that a compound is released rapidly from an emulsion formulation when its calculated logP (CLogP) is less than 7. However, if its CLogP is greater than 8, then there is a greater delivery to the liver and spleen as a result of removal of intact oil droplets and compound.

Excipient toxicity

As the need to deliver poorly soluble compounds rises, the use of excipients that can exert their own toxicological effects has increased. All the excipients that are described in this chapter and in their accompanying datasheets (in the Annex) will cause toxic effects when dosed at a high enough level, as evidenced by their LD_{50}. Generally the maximum dose of an excipient should not exceed 1/4 of the LD_{50} (Bartsch et al. 1976), but this is a rather crude method of selecting maximum doses as some excipients will exhibit a steeper dose/toxic effect response than others.

This section describes some of the excipient toxicities that have been observed around frequently used concentrations and doses. The toxicity is governed by many factors that must be considered, such as concentration in the formulation, rate of delivery, duration of use, and overall dose.

Tonicity modifiers

Tonicity modifiers are essential in simple aqueous formulations and usually considered innocuous from a toxicological perspective. However, the modifying agent must be chosen carefully, as there may still be consequences for the animal being infused. A 0.9% w/v sodium chloride solution places little burden on an animal if infused within normal volume limits, as it is not metabolised, is rapidly distributed into the extracellular fluid, and is voided in the urine in the same volume infused. However, agents that are metabolised, such as glucose, increase the burden on an animal and may induce physiological changes, such as increased arterial pressure in dogs (Brands et al. 2001), and increase the incidence of opportunistic infections in long-term studies (Salauze and Cave 1995). Agents such as mannitol reduce renal medullary bloodflow (Liss et al. 1996).

Cosolvents

Commonly used cosolvents in parenteral formulation are known to have numerous toxic effects when dosed parenterally. These effects normally occur in a dose-dependent manner and vary in severity. Table 4.5 summarises a small number of findings specific to the cosolvents listed. Cosolvents also have the disadvantage that they may indirectly cause

Table 4.5 Potential Toxic Effects of Cosolvents

Cosolvent	Effect	Species or system	Drug used in reference	References
DMSO	Osmotic diuresis	Human	N/A	Gunn and Acomb 1986
DMSO	Haemolysis, white cell sticking, and fibrinogen precipitation at concentration >50%	Human blood	N/A	Johnson et al. 1966
Ethanol	Long term administration causes decreased albumin synthesis	Human	N/A	Rothschild et al. 1975
Propylene glycol	Nephrotoxicity, cardiac depression	Human	N/A	Hayman et al. 2003; Yaucher et al. 2003; Levy et al. 1995
PEGs	Acute tubular necrosis	Human	N/A	Laine et al. 1995
Glycofurol	Significant signs of toxicity and alteration in the pharmacologic actions of barbiturates	Rat	Barbiturates	Yasaka et al. 1978

toxicity through precipitation of the administered compound owing to the rapid drop in solubility of compound upon dilution in plasma (see section on precipitation).

Surfactants

Surfactants have the highest potential for toxicity, in part due to the same mechanisms that solubilise compounds, by affecting biological membranes, interacting with plasma proteins, and altering the pharmacokinetics of co-administered compounds. This can limit their use, though they remain a useful tool to enable the delivery of challenging compounds. The surfactants discussed in this section consist of Cremophor EL and Tween 80 (polysorbate 80), as they have widely reported toxic effects. The other parenterally acceptable surfactants, such as Solutol HS 15 and Pluronic® F68 (poloxamer 188), do not have as many reported cases of toxicity when dosed well below their LD_{50} and appear to be safer.

Cremophor EL and Tween 80 are widely used in clinical and preclinical parenteral formulation. Their pharmacokinetics may contribute to the toxic effects of a formulation. Their half-lives in humans vary from 10 to 140 hours for Cremophor EL to less than 30 minutes for Tween 80 (Bittner and Mountfield 2002). Cremophor EL has a low volume of distribution in the range of the central blood compartment.

Acute hypersensitivity reactions are seen in a clinical setting, characterised by dyspnea, flushing, rash, and generalised urticaria (Onetto et al. 1995). Pre-clinically, Lorenze et al. (1977 and 1982) reported that Cremophor EL caused significant histamine release in dogs mainly by its minor free fatty acid constituents such as oleic acid. Oleic acid is also present in Tween 80 and may be responsible for the hypersensitivity reactions that are seen in dogs (Essayan et al. 1996).

Cremophor EL and Tween 80 have been implicated in neuropathy (ten Tije et al. 2003). In both cases, it is difficult to rule out compound-related effects, but the evidence is stronger for Cremophor EL. Neuropathy is observed in ~25% of patients who receive cyclosporine intravenously (de Groen et al. 1987), but this effect disappears with the oral formulations where Cremophor EL would not be absorbed from the GI tract (ten Tije et al. 2003).

Lipoprotein alterations are also noted with Cremophor EL, where it was found to alter the buoyant density of high-density lipoproteins (Kongshaug et al. 1991) and potentially cause hyperlipidaemia (Shimomura et al. 1998).

Tween 80 has also been implicated in acute hepatitis (Rhodes et al. 1993) when administered with parenteral amiodarone in a clinical setting, but as with many other studies with surfactants it is impossible to exclude compound effects.

Cyclodextrins

A variety of cyclodextrins, including α-cyclodextrin and γ-cyclodextrin, exhibit renal toxicity and disrupt biological membranes (Stella and He 2008). However, the two cyclodextrins that this chapter focuses on (HP-β-CD and SBE-β-CD) appear to be much safer. There are at least 22 commercial cyclodextrin-based pharmaceutical products, of which 14 are marketed in Europe, 7 in Japan, and 5 in the US, with more CD-based products awaiting Regulatory approval.

This section highlights some of the key toxicological findings that relate to the use of HP-β-CD and SBE-β-CD. Most of the information here is from reviews such as Gould and Scott (2005) and Thompson (1997).

Renal toxicity is attributed to kidney reabsorption of unmodified cyclodextrins (Rajewski et al. 1995). HP-β-CD and SBE-β-CD were produced to increase the hydrophilic nature and reduce this effect. The mechanism of injury is thought to occur through complexation with cholesterol, which precipitates as intracellular needle-like crystals that cause kidney damage (Frijlink et al. 1991).

HP-β-CD and SBE-β-CD cause vacuolation of the proximal tubular epithelium. The no-observable-effect-levels in rats are >50mg/kg daily for 3 months (Gould and Scott 2005) and 80mg/kg daily for 1 month (Thompson 1997). However, in contrast to β-CD, this effect is reversible and does not affect renal function (Stella and He 1998).

The toxicity of HP-β-CD has been extensively investigated and was found to cause some minor clinical observations, as well as biochemical and histopathology changes when dosed intravenously. The target organs were the lungs (increased macrophage infiltration), liver, and kidney. Each effect had a no-effect level and was reversible (Gould and Scott 2005).

Tumours of the exocrine pancreas were found at all dose levels after 12 of months of daily dosing in rats, and were considered a consequence of bile sequestration causing increased cholecystokinin (CCK) release and cell proliferation. This was reported to be a rat-specific mechanism and has not been observed in other species (Gould and Scott 2005).

Some effects on the haemolytic potential of cyclodextrins have been published showing that HP-β-CD can cause haemolysis at high concentrations (Rajewski et al. 1995) compared to SBE-β-CD having negligible effects. This is put forward as a marker of their membrane disruption potential, possibly due to complexation with membrane cholesterol. However, in most other preclinical and clinical studies there are no other safety concerns reported for HP-β-CD and SBE-β-CD when used within guideline concentrations.

Strategies for dealing with injection site reactions and haemolysis

Injection site reactions are a key concern for intravenous infusions. They can present themselves in a number of different ways, such as pain, phlebitis, thrombophlebitis, infiltration, extravasation, local infection, haematoma, and thrombosis. Phlebitis and related conditions, considered in this section, manifest as pain, erythema, swelling, and localised temperature increasing over a period of hours or days (Sweetana and Akers 1996).

The act of puncturing the epidermis can cause an injection site reaction even before the formulation is administered, so some reactions are to be expected over the course of a study. Predictive models of injection site irritation are useful but are imperfect (Sweetana and Akers 1996), so it is important to understand its potential root causes. Figure 4.10 is a map of potential root causes of injection site reactions, with arrows showing some of the many and varied interdependencies.

There are a large number of potential causes of injection site reactions. Even though the causes will differ from study-to-study, some appear more frequently in literature. Whilst every possibility must be considered when an injection site reaction is encountered in practice, the bolded text in Figure 4.10 highlights some of the more influential causes described below:

- pH, buffer strength, osmolality, and duration of infusion
- Precipitation
- Compound irritation
- Choice of vein

pH, buffer strength, osmolality and duration of infusion

The pH of a formulation is often thought to be the cause of injection site reactions during infusion studies. The most common recommendation to ensure biocompatibility is to infuse at a pH between 4 and 8 (Sweetana and Akers 1996). However, it is frequently necessary to use a pH outside this range to ensure complete dissolution of a compound. Sweetana and Akers (1996) have a listing of marketed parenteral products administered between pH 2 and 11. In addition, Figure 4.11 shows the frequency of parenteral products that are dosed at a range of volumes and pHs.

That so many infusion products dosed at extremes of pH and volume are marketed indicates that pH alone may not be solely responsible for injection site reactions. This is consistent with the findings of Simamora et al. (1995). Buffers of varying pH and low buffer capacity were injected

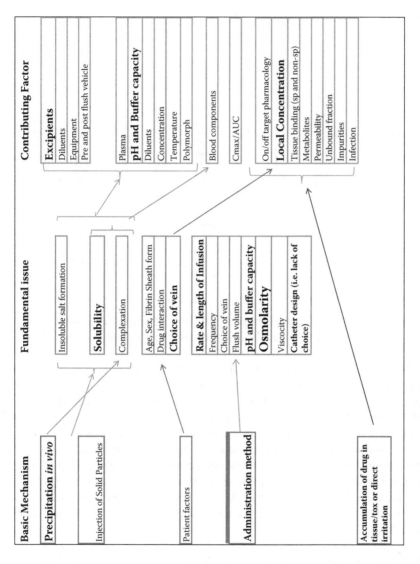

Figure 4.10 (See colour insert.) Map of injection site reaction root causes.

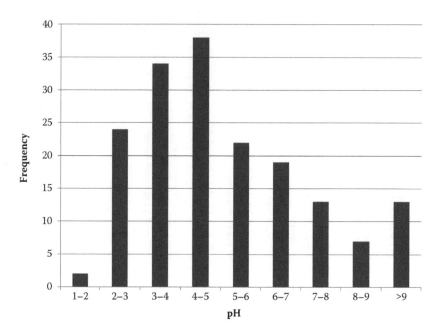

Figure 4.11 Dose volumes and pH ranges for a variety of commercial intravenous products.

in the rabbit ear model, and those within the pH range 2–11 were found not to cause any sign of phlebitis, whereas higher pHs (12 and 13) did produce signs of phlebitis.

Based on these findings, it is likely that factors other than pH alone contribute to injection site reactions. Kuwahara et al. (1996) demonstrated that the buffer capacity and pH influenced the induction of phlebitis in the rabbit ear model. Solutions with different buffer capacities were infused into the ear veins of 6 rabbits for 6 hours at 10 mL/kg/h. A pH 4.0, 10% glucose solution of low buffer capacity caused negligible changes, but when the buffer capacity was moderate (pH 4.3) phlebotic changes were observed in all 6 rabbits. Increasing the buffer capacity further (pH 4.4) increased the degree of phlebitis. A 10% glucose solution with a pH of 5.4 and of moderate buffer capacity caused slight phlebotic changes in half the rabbits.

It is likely that higher buffer capacities overwhelm an animal's ability to buffer the formulations back to physiological pH and that the closer the formulation is to physiological pH, the less severe the effects are.

In a different study, Kuwahara et al. (1998) examined the relationship between osmolality and duration of infusion. Nutrient solutions of 539–917 mOsm/kg were infused into rabbit ear veins for 8, 12, or 24 hours.

Histopathology performed on the veins found that higher osmolality solutions caused some phlebotic changes, such as loss of venous endothelial cells and inflammatory cell infiltration, and the lowest osmolality solution caused few changes. Infusion of 120 mL/kg of 814 mOsm/kg solution caused phlebitis at 5 or 10 kg^{-1} h^{-1}, but the same volume produced negligible effects at 15 mL kg^{-1} h^{-1}. It was postulated that this was caused by the shorter infusion time at the higher rate.

Even though the ideal osmolality range is 280–290 mOsm/kg, it is clear from these studies that higher values can be tolerated and other factors such as infusion time (and therefore rate of dilution) affect the osmolalities that can be tolerated. This is consistent with the recommendation of the Infusion Nurses Society, who recommend values for maximum osmolality ranging from below 500 to above 900, determined by the choice of vein and flow rates (i.e. dilution rates) in those veins.

Osmolality can be calculated based on ideal behaviour. For example, it is known that 0.9% sodium chloride is iso-osmotic, cosolvents such as ethanol, propylene glycol and DMSO have iso-osmotic concentrations of 2% v/v, and 28% w/v HP-β-CD is iso-osmotic. However, not all solutes behave ideally, and methods such as freezing point depression are required to assess osmolality.

An understanding of how these parameters affect the local tolerability of a formulation should allow for a design that minimises its irritant effect. A simple rule is to use a formulation as close to physiological parameters as possible. However, the potential interplay between pH, buffer strength, osmolality, and duration of infusion make predictions of local tolerability somewhat difficult. There are even conflicting reports in the literature that demonstrate the complexity of these effects. Ward and Yalkowsky (1993) studied the infusion of an irritating compound (amiodarone) in the rabbit ear model at flow rates from 0.02 mL to 3 mL min^{-1}. Interestingly, minimal phlebitis was observed at the slowest and fastest infusion rates. These results may be affected by precipitation of the compound upon administration at higher rates (see next section).

Precipitation

Precipitation of compound is one of the major causes of injection site reactions. Myrdal et al. (1995) demonstrated that by using a weak buffer to prevent precipitation of levemopamil in the rabbit ear model, phlebitis could be avoided compared to using a precipitating formulation. Simamora et al. (1996) had similar findings with dexaverapamil.

Precipitation *in vivo* is caused primarily by a change in the conditions leading to supersaturation (a concentration above the saturated solubility) of compound in solution. The conditions that might cause this are summarised in Figure 4.12, reproduced from Yalkowsky (1999), which depicts

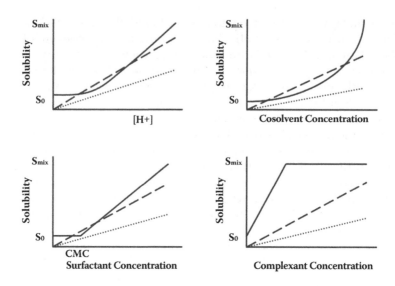

Figure 4.12 Precipitation of dissolved compound on dilution in different systems.

compounds dissolved in the various formulation systems describe earlier in this chapter. The two dotted lines represent the dilution from two different concentrations of compound to infinite dilution (moving from right to left). The pH-adjusted, cosolvent, and surfactant formulations go through regions where the higher concentration of compound is supersaturated with respect to its conditions. In the complexed formulation, the complex is maintained throughout dilution and no supersaturation is observed.

Passing through a region of supersaturation on dilution increases the risk of precipitation. In practice this does not always happen, as precipitation is time dependent and may not have time to occur before further dilution in blood occurs. Protein binding also takes place to remove free compound.

Precipitation testing

In vitro precipitation testing can indicate the likelihood of precipitation upon IV administration. Li et al. (1998) tested three *in vitro* methods—static serial dilution, dynamic injection and drop-wise additions—and found them all to be complementary. One of the simpler methods used was serial dilution. The formulation is sequentially diluted in a one-to-one ratio with isotonic Sorensen's phosphate buffer: 1:1 to 1:2 to 1:4 to 1:8 to 1:16 to 1:32 to 1:64, etc. Isotonic Sorensen's phosphate buffer was chosen because it mimics physiological pH (7.4) and buffer capacity (~0.036 M). The advantages of the serial method over the others were that it was most effective in quantifying the amount of precipitation and was more descriptive of the formation and re-dissolution of the precipitate. Animal plasma can also be use as a more physiologically relevant medium, but buffers offer 'worst case' conditions. The dynamic injection test was more physiologically relevant and involved 1 mL of formulation injected at several constant rates (0.2, 0.5, 1.0, 2.0, 5.0, 10.0, and 20.0 mL/min) into isotonic Sorensen's phosphate buffer that was pumped at 20 mL/min. Particles were detected in a flow cell within a spectrophotometer.

Precipitation avoidance
If the mechanism of precipitation is known, then it may be possible to take steps to reduce or avoid it upon intravenous administration. A number of strategies are described here.

Buffer Capacity. If precipitation is found to occur from rapid buffering between formulation and physiological pH, then it may be possible to adjust the buffer capacity of the formulation to resist pH change for long enough for further *in vivo* dilution to occur, as found by Myrdal et al. (1995) and Simamora et al. (1996). Care must be taken not to over-buffer an animal, as this may lead to deleterious changes in blood chemistry or increased irritation.

Complexing agents. Figure 4.12 shows that cyclodextrins can be effective at preventing precipitation by forming a stable complex. If solution is achieved in the formulation, precipitation will not occur on dilution as long as the complexed compound is un-ionised and forms a 1:1 complex. In practice, these conditions are rarely met, as pH adjustment (ionisation) is often used with cyclodextrins. In these cases there is still a precipitation risk, but cyclodextrins still provide enough protection to warrant their use.

Use of a surfactant or formulate as an emulsion. If a compound is amenable to formulation in a surfactant or emulsion system, then this can be an effective way to protect the compound from the aqueous environment at the instant of administration. This potentially allows for a greater level of dispersion of the micelles/droplets before the compound is released into plasma below its saturation solubility. Care must still be taken because of the potential for altered pharmacokinetics, as described earlier in this chapter.

Increased infusion time. Increasing the infusion time while using the same infusion volume reduces the infusion rate and effectively increases the dilution rate. This provides an opportunity for the compound to be rapidly diluted to a concentration below its saturated solubility in plasma. This is the case in studies conducted by Ward and Yalkowsky (1993). However, increasing infusion time may introduce other factors, such as prolonged contact with an irritant compound or high osmolality, that may also cause an injection site reaction.

Nanosuspensions. If a compound is amenable to a nanosuspension formulation, then infusion of solids of a controlled particle size may reduce its phlebitis potential. This would generally be considered a last resort because of the development time and cost, repeat PK studies, and the potential influence on disposition a nanosuspension may have.

Compound irritation

Frequently, a formulation, or the method used to administer it, is not the cause of an injection site reaction that is occurring. In these situations, the conclusion is that the compound is an irritant. This may be caused by compound being absorbed and concentrating into local tissues or by direct irritation of the venous endothelial cells. The strategies used to deal with this target a reduction of the local compound concentration.

Reduced infusion time
As Ward and Yalkowsky (1993) found, reducing the infusion time (i.e. increasing the rate) decreased a formulation's phlebitic potential. The reduced contact time with high concentrations of the compound should then lead to a reduced uptake into local tissue or endothelial cells. This effect is likely to be dependent on the compound and the interplay between the rate of absorption into tissues and the rate of infusion. It may also increase the potential for precipitation.

Altering the ionisation state
The ionisation state of a compound can affect its membrane permeability. Using pH adjustment or a buffer in a formulation may decrease its permeability to prevent local accumulation and reduce injection site reactions.

Emulsions, surfactants and complexing agents
Another approach is to protect tissues close to the site of injection by reducing the free compound concentration. This could be achieved with surfactants to form micelles, cyclodextrins (Stella and He 2008), or emulsions. Cantwell et al. (2006) showed that an emulsion formulation of a phlebitis-inducing anti-cancer compound (vinorelbine) significantly reduced the potential for phlebitis.

Choice of vein

The choice of vein can also have a significant impact on the incidence of injection site reactions. Larger vessels with higher flow rates will have a lower incidence of phlebitis. This is because the larger diameter causes less occlusion to blood flow when a cannula is inserted, allowing more rapid blood flow and dilution. The compound concentration at the vessel wall is reduced compared with a smaller vessel owing to laminar flow. A concentration gradient develops from the centre of the vessel to the walls from the way the plume of formulation leaves the cannula tip (Yalkowsky et al. 1998).

Haemolysis

Haemolysis occurs when red blood cells rupture through disruption of their membranes, causing a release of their contents. The release of haemoglobin can be problematic if it is more than the body can remove and can lead to clogged renal tubules and affect kidney and liver function (Krzyzaniak et al. 1996).

Excipients and compounds as well as the infusion of hypo-osmotic solutions have the potential to cause haemolysis from osmotic swelling and rupture. Parenteral excipients known to cause haemolysis include HP-β-CD (Rajewski et al. 1995) and propylene glycol (Fu et al. 1987).

A number of methods to assess the haemolytic potential of formulation have been reviewed by Yalkowski et al. (1998). The methods all involve exposing a blood sample to the formulation for a defined time and measuring the level of haemoglobin leakage into plasma. Each method reviewed gave different results, with the contact time being an important parameter. A one-second contact time was found to be most predictive.

The strategies that could be used to reduce compound-induced haemolysis are primarily the same as those to reduce irritation, as they may have the same cause, that is, membrane disruption. Excipient-induced haemolysis is typically ameliorated by slowing down the infusion rate, switching to an alternative, or adding protective excipients. For example, the haemolytic effects of propylene glycol can be reduced by the addition of PEG400 (Fu et al. 1987).

Strategies for dealing with poor stability

The majority of unstable compounds experience their most rapid degradation whilst in the solution state. This can make formulating for intravenous infusion studies challenging for inherently unstable molecules. However, there are a number of strategies used to mitigate these issues that depend upon the mechanism of degradation.

Simple general good practice steps that should be taken to minimise degradation include fresh preparation where possible, protecting the formulation from light during storage and administration, refrigeration of the formulation where warranted, and warming to room temperature for administration for the minimum amount of time. However, these steps may not be enough to prevent unacceptable levels of degradation.

Removal of water

As hydrolysis is the most common cause of instability, water removal can often be the only effective strategy for preventing hydrolysis of a compound. This is less of a problem for formulations that are prepared and filter sterilised (see Sterility section) just prior to administration, but can be a problem where a formulation needs to be stored for an extended period in a sterile state. Freeze-drying (lyophilisation) can be very effective in these circumstances, but a compound must be amenable to it and is rarely used for pre-clinical infusion studies. Jennings (1999) contains a details description of the process.

Another seldom-used method of reducing a compound's contact with water is to formulate in a non-aqueous vehicle, such as an emulsion or a nano-suspension. Whilst these are possible, they could introduce other concerns over toxicity or changes in PK (see Unwanted Formulation Effects section).

Additives

Certain additives may reduce the rate of chemical degradation, depending on its mechanism. Where oxidation is an issue, the use of an antioxidant may be warranted. There are a number of parenterally acceptable antioxidants, such as sodium bisulfite, sodium metabisulfite, and ascorbic acid, that work by being preferentially oxidised in place of the compound. However, the risk of side reactions of these oxidised products with the compound must be studied before their use.

Another strategy for reducing oxidation is to purge the water used to prepare the formulation and/or maintain the solution under an inert gas such as argon or nitrogen.

Some oxidation reactions are catalysed by trace metals left over from the synthesis of compounds. In these cases an additive such as edetate disodium or citric acid have the ability to chelate trace metal ions and inhibit their catalysis of oxidation.

Sterility

Ideally, it is desirable to deliver sterile, endotoxin-free formulations in all studies. However, this is not always practical or required. For example, early single dose, high throughput PK studies probably do not require any more than 'clean' preparation to ensure that the endpoint of these studies is achieved. However, it is far more important to assure the sterility of toxicity studies to ensure that the effects of a compound are not confounded by the response to an infection or endotoxin.

As for products intended for human use, it is possible to calculate an endotoxin limit for parenteral formulations. Malaya and Singh (2007) describe some of these limits for preclinical studies.

At a minimum the following steps must be taken to ensure sterility (Shah and Agnihotri 2011):

- Preparation should be conducted using aseptic techniques in sterile bio-safety cabinet or laminar flow hood.
- All surfaces in preparation areas should be wiped down with hydrogen peroxide solution.
- Glassware should be de-pyrogenated.
- Excipients should be low endotoxin.

Pre-clinical studies that require the use of aqueous solution formulations usually utilise filter sterilisation as a last step before aseptic filling into sterile, pyrogen-free glassware or syringe for administration. It is important to select a compatible filter for the formulation during formulation development. Filter membrane materials include polyvinylidene difluoride (PVDF) and polytetrafluorethylene (PTFE).

Generally, 0.22 μm filters are used to ensure sterility. However, it may not be possible to sterilise using the filter sterilisation of some of the formulation approaches described in this chapter. Emulsions and nanosuspensions may not be able to successfully pass through filters because of the presence of multiple phases (solids or emulsion droplets). High cosolvent levels in a formulation may not be chemically compatible with any filter type. Achieving sterility assurance can be challenging under these circumstances, and it may be necessary to employ other methods such as autoclaving and gamma irradiation. However, not all compounds will be amenable to these methods, because of stability issues, and this may necessitate aseptic and sterile compound and formulation manufacture that can increase costs and timelines significantly.

Conclusion

Pre-clinical formulation development is critical to understanding all the aspects of a compound's *in vivo* performance. It is of vital importance to understand the properties of the system administered in an intravenous formulation—it involves more than just the compound.

The challenges associated with pre-clinical intravenous formulation development are not only multi-factorial but also interact in complex and non-linear ways that are sometimes difficult to predict. Success depends on an understanding of compound properties, the objectives (PK, late stage safety, early stage safety, efficacy, etc.) of the study, and the species of choice so that appropriate excipients and concentrations can be chosen.

In all circumstances, it is best to use the simplest formulation possible, as they are easier to understand, prepare, and troubleshoot if failure occurs. This chapter and the reference sources it contains should be used as a starting point in developing and understanding the impact a formulation has on an intravenous infusion study.

Common Excipients and Vehicles data sheets are located in the Annex.

References

Albini A and Fashini E. 1998. *Drugs: Photochemistry and Photostability*. Cambridge, UK: The Royal Society of Chemistry.

Azmin MN, Stuart JFB and Florence AT. 1985. The distribution and elimination of methotrexate in mouse blood and brain after concurrent administration of polysorbate 80. *Cancer Chemotherapy Pharmacology* 14: 238–242.

Bardelmeijer HA, Ouwehand M, Malingre MM, Schellens JH, Beijnen JH and van Tellingen O. 2002. Entrapment by Cremophor® EL decreases the absorption of paclitaxel from the gut. *Cancer Chemotherapy Pharmacology* 49: 119–125.

Bartsch W, Sponer G, Dietmann K and Fuchs G. 1976. Acute toxicity of various solvents in the mouse and rat: LD50 of ethanol, diethylacetamide, dimethylformamide, dimethyl sulfoxide, glycerin, n-methylpyrrolidone, polyethylene glycol 400, 1,2-propanediol and Tween® 20. *Arnzneimittelforschung* 26: 1581–1583.

Bittner B and Mountfield RJ. 2002. Intravenous administration of poorly soluble new drug entities in early drug discovery: The potential impact of formulation on pharmacokinetic parameters. *Current Opinion in Drug Discovery and Development* 5: 59–71.

Box KJ and Comer JEA. 2008. Using measured pKa, logP and solubility to investigate supersaturation and predict BCS class. *Current Drug Metabolism*, 9, 869–878.

Brands MW, Hailman AE and Fitzgerald SM. 2001. Long-term glucose infusion increases arterial pressure in dogs with cyclooxygenase-2 inhibition. *Hypertension* 37: 733–738.

Brink JJ and Stein DG. 1967. Pemoline levels in brain: Enhancement by dimethyl sulfoxide. *Science* 158: 1479–1480.

Bristow S, Shekunov T and Shekunov B. 2001. Analysis of the supersaturation and precipitation process with supercritical CO_2. *Journal of Supercritical Fluids* 21: 257–282.

Buggins TR, Dickinson PA and Taylor G. 2007. The effects of pharmaceutical excipients on drug disposition. *Advanced Drug Delivery Reviews* 59: 1482–1503.

Cantwell MJ, Robbins JM and Chen AX. 2006. A novel emulsion formulation of vinorelbine attenuates venous toxicity while maintaining antitumor efficacy. *Proceedings of the American Association for Cancer Research* 47: abs.

Chauret N, Gauthier A and Nicoll-Griffith DA. 1998. Effect of common organic solvents on *in vitro* cytochrome P450-mediated metabolic activities in human liver microsomes. *Drug Metabolism and Disposition* 26: 1–4.

Connors KA, Amidon GA and Stella VJ. 1986. *Chemical Stability of Pharmaceuticals: A Handbook for Pharmacists*, 2nd edition. New York: John Wiley and Sons.

de Groen PC, Aksamit AJ, Rakela J, Forbes GS and Krom RA. 1987. Central nervous system toxicity after liver transplantation: The role of cyclosporine and cholesterol. *New England Journal of Medicine* 317: 861–866.

Diehl KH, Hull R, Morton D, Pfister R, Rabemampianina Y, Smith D, Vidal J-M and van de Vorstenbosch C. 2001. A good practice guide to the administration of substances and removal of blood, including routes and volumes. *Journal of Applied Toxicology* 21: 15–23.

Douroumis D and Fahr A. 2007. Stable carbamazepine colloidal systems using the cosolvent technique. *European Journal of Pharmaceutical Science* 30: 367–374.

Ellis AG, Crinis NA and Webster LK. 1996. Inhibition of etoposide elimination in the isolated perfused rat liver by Cremophor EL and Tween 80. *Cancer Chemotherapy and Pharmacology* 38: 81–87.

Essayan DM, Kagey-Sobotka A, Colarusso PJ, Lichtenstein LM, Ozols RF and King ED. 1996. Successful parenteral desensitization to paclitaxel. *Journal of Allergy Clinical Immunology* 97: 42–46.

Frijlink HW, Franssen EJF, Eissens AC, Oosting R, Lerk CF and Meijer DKF. 1991. The effects of cyclodextrins on the disposition of intravenously injected drugs in the rat. *Pharmaceutical Research* 8: 380–384.

Fu RC, Lidgate DM, Whatley JL and McCullough T. 1987. The biocompatibility of parenteral vehicles: *In vitro/in vivo* screening comparison and the effect of excipients on hemolysis. *Journal of Parenteral Science and Technology* 41: 164–168.

Gibson M. 2001. *Pharmaceutical Preformulation and Formulation*. London: Taylor and Francis.

Gould S and Scott RC. 2005. 2-hydroxypropyl-β-cyclodextrin (HP-β-CD): A toxicology review. *Food Chemical Toxicology* 43: 1451–1459.

Gould PL. 1986. Salt selection for basic drugs. *International Journal of Pharmaceutics* 33: 201–217.

Gunn IR and Acomb C. 1986. High plasma osmolality following intravenous dimethyl sulfoxide in the treatment of postoperative hemiplegia. *Journal of Neurology, Neurosurgery and Psychiatry* 49: 961–962.

Hansrani PK, Davis SS and Groves MJ. 1983. The preparation and properties of sterile intravenous emulsions. *J. Parenter. Sci. Technol.* 37: 145–150.

Hayman M, Seidl EC, Ali M and Malik K. 2003. Acute tubular necrosis associated with propylene glycol from concomitant administration of intravenous lorazepam and trimethoprim-sulfamethoxazole. *Pharmacotherapy* 23: 1190–1194.

Higuchi T and Connors KA. 1965. Phase-solubility techniques. *Advances in Analytical Chemistry and Instrumentation* 4: 117–122.

Jennings TA. 1999. *Lyophilization: Introduction and Basic Principles*. Denver, Col.: Interpharm Press.

Jin M, Shimada T, Yokogawa K, Nomura M, Mizuhara Y, Furukawa H, Ishizaki J and Miyamoto K. 2005. CremophorEL releases cyclosporin A adsorbed on blood cells and blood vessels, and increases apparent plasma concentration of cyclosporin A. *International Journal of Pharmaceutics* 293: 137–144.

Johnson JH, Baker RR and Wood S. 1966. Effects of DMSO on blood and vascular endothelium. In: *Fourth European Conference on Microcirculation*, Harders H (ed.). Basel: Karger.

Kamath AV, Chang M, Lee FY, Zhang Y and Marathe PH. 2005. Preclinical pharmacokinetics and oral bioavailability of BMS-310705, a novel epothilone B analog. *Cancer Chemotherapy and Pharmacology* 56: 145–153.

Kambam JR, Franks JJ, Janicki PK, Mets B, van der Watt M and Hickman R. 1994. Alcohol pretreatment alters the metabolic pattern and accelerates cocaine metabolism in pigs. *Drug Alcohol Dependence* 36: 9–13.

Kassell NF, Sprowell JA, Boarini DJ and Olin JJ. 1983. Effect of dimethyl sulfoxide on the cerebral and systemic circulations of the dog. *Neurosurgery* 12: 24–28.

Kaus L. 1998. Buffers and buffering agents. In: *Encyclopedia of Pharmaceutical Technology*, Vol. 2, Swarbrick J and Boylan JC (eds), pp. 213–231. New York: Marcel Dekker.

Keck CM and Müller RH. 2006. Drug nanocrystals of poorly soluble drugs produced by high pressure homogenisation. *European Journal of Pharmaceutics* 62: 3–16.

Kongshaug M, Cheng LS, Moan J and Rimington C. 1991. Interaction of Cremophor EL with human serum. *International Journal of Biochemistry* 23: 473–478.

Kontir DM, Glance CA, Clolby HD and Miles PR. 1986. Effects of organic solvent vehicles on benzo(a)pyrene metabolism in rabbit lung microsomes. *Biochemistry and Pharmacology* 35: 2569–2575.

Krzyzaniak JF, Raymond DM and Yalkowsky SH. 1996. Lysis of human red blood cells 1: Effect of contact time on water induced hemolysis. *PDA Journal of Pharmaceutical Sciences and Technology* 50: 223–226.

Kumar L. 2007. An overview of automated systems relevant in pharmaceutical salt screening. *Drug Discovery Today* 12: 1046–1053.

Kurkov SV, Loftsson T, Messner M and Madden D. 2010. Parenteral delivery of HPβCD: Effects on drug-HSA binding. *AAPS Pharmaceutical Science and Technology* 11: 1152–1158.

Kurlansky PA, Sadeghi AM, Michler RE, Coppey LJ, Re LP, Thomas WG, Smith CR, Reemtsma K and Rose EA. 1986. Role of the carrier solution in cyclosporine pharmacokinetics in the baboon. *Journal of Heart Transplant* 5: 312–316.

Kuwahara T, Asanami S and Kubo S. 1998. Experimental infusion phlebitis: Tolerance osmolality of peripheral venous endothelial cells. *Nutrition* 14: 496–501.

Kuwahara T, Asanami S, Tamura T and Kubo S. 1996. Experimental infusion phlebitis: Importance of titratable acidity on phlebitic potential of infusion solution. *Clinical Nutrition* 15: 129–132.

Laine GA, Hossain SM, Solis RT and Adams SC. 1995. Polyethylene glycol nephrotoxicity secondary to prolonged high-dose intravenous lorazepam. *Annals of Pharmacotherapy* 29: 1110–1114.

Levy ML, Aranda M, Zelman V and Giannotta S L. 1995. Propylene glycol toxicity following continuous etomidate infusion for the control of refractory cerebral edema. *Neurosurgery* 37: 363–371.

Li P and Zhao L. 2007. Developing early formulations: Practice and perspective. *International Journal of Pharmaceutics* 341: 1–19.

Li P, Vishnuvajjala R, Tabibi SE and Yalkowsky SH. 1998. Evaluation of *in vitro* precipitation methods. *Journal of Pharmaceutical Sciences* 82: 162–169.

Lipinski CA. 2000. Drug-like properties and the causes of poor solubility and poor permeability. *Journal of Pharmacology and Toxicology Methods* 44: 235–249.

Liss P, Nygren A, Olsson U, Ulfendahl HR and Erikson U. 1996. Effects of contrast media and mannitol on renal medullary blood flow and red cell aggregation in the rat kidney. *Kidney International* 49: 1268–1275.

Loos WJ, Baker SD, Verweij J, Boonstra JG and Sparreboom A. 2003. Clinical pharmacokinetics of unbound docetaxel: Role of polysorbate 80 and serum proteins. *Clinical Pharmacology and Therapeutics* 74: 364–371.

Lorenze W, Schmal A, Schult H, Ohmann C, Weber D, Kapp B, Lüben L and Doenicke A. 1982. Histamine release and hypotensive reactions in dogs by solubilizing agents and fatty acids: Analysis of various components in Cremophor EL and development of a compound with reduced toxicity. *Agents Actions* 12: 64–80.

Lorenze W, Reimann HJ, Schmal A, Dormann P and Schwarz B. 1977. Histamine release in dogs by Cremophor EL and its derivates: Oxethylated oleic acid is the most effective constituent. *Agents Actions* 7: 63–67.

Loscher W, Honack D, Richter A, Schulz HU, Schurer M, Dusing R and Brewster ME. 1995. New injectable aqueous carbamazepine solution through complexing with 2-hydroxypropyl-beta-cyclodextrin: Tolerability and pharmacokinetics after intravenous injection in comparison to a glycofurol-based formulation. *Epilepsia* 36: 255–261.

Malingre MM, Schellens JH, van Tellingen O, Ouwehand M, Bardelmeijer HA, Rosing H, Koopman FJ, Schot ME, Ten Bokkel Huinink WW and Beijnen JH. 2001. The co-solvent Cremophor EL limits absorption of orally administered paclitaxel in cancer patients. *British Journal of Cancer* 85: 1472–1477.

Malyala P and Singh M. 2007. Endotoxin limits in formulations for preclinical research. *Journal of Pharmaceutical Sciences* 97: 2041–2044.

Merisko-Liversidge E, Liversidge GG and Cooper ER. 2003. Nanosizing: A formulation approach for poorly-water-soluble compounds. *European Journal of Pharmaceutical Sciences* 18: 113–120.

Mosher GL and Thompson DO. 2002. Complexation and cyclodextrins. *Encyclopedia of Pharmaceutical Technology* 2nd ed., J. Swarbrick and J. C. Boylan (eds), pp. 531–558. New York: Marcel Dekker.

Mouton JW, Van Peer A, De Beule K, Donnelly JP and Soons PA. 2006. Pharmacokinetics of itraconazole and hydroxyitraconazole in healthy subjects after single and multiple doses of a novel formulation. *Antimicrobial Agents and Chemotherapy* 50: 4096–4102.

Müller RH, Benita S and Böhm BHL. 1998. *Emulsions and Nanosuspensions for the Formulation of Poorly Soluble Drugs.* Stuttgart, Germany: Medpharm Scientific Publishers.

Myrdal PB, Simamora P, Surakitbanharn Y and Yalkowsky SH. 1995. Studies in phlebitis. VII: *in vitro* and *in vivo* evaluation of pH-solubilized levemopamil. *Journal of Pharmaceutical Sciences* 84: 849–852.

Neervannan S. 2006. Preclinical formulation for discovery and toxicology: Physicochemical challenges. *Expert Opinion on Drug Metabolism and Toxicology* 2: 715–731.

Newman AW and Byrn SR. 2003. Solid-state analysis of the active pharmaceutical ingredient in drug products. *Drug Discovery Today* 8: 898.

Oguma T and Levy G. 1981. Acute effect of ethanol on hepatic first-pass elimination of propoxyphene in rats. *Journal of Pharmacology and Experimental Therapeutics* 219: 7–13.

Onetto N, Dougan M, Hellmann S, Gustafson N, Burroughs J, Florczyk A, Canetta R and Rozenweig M. 1995. Safety profile. In: *Paclitaxel in Cancer Treatment*, McGuire WP and Rowinsky EK (eds), pp. 21–149. New York: Marcel Dekker.

Parker RB and Laizure C. 2010. Effect of ethanol on oral cocaine pharmacokinetics. *Drug Metabolism and Disposition* 38: 317–322.

Perry CS, Charman SA, Prankerd RJ, Chiu FC, Scanlon MJ, Chalmers D, and Charman WN. 2006. The binding interaction of syntheticozonide antimalarials with natural and modified β-cyclodextrins. *Journal of Pharmaceutical Science* 95: 146–158.

Pestel S, Martin HJ, Maier GM and Guth B. 2006. Effect of commonly used vehicles on gastrointestinal, renal, and liver function in rats. *Journal of Pharmacology and Toxicology Methods* 54: 200–214.

Rabinow B, Kipp J, Papadopoulos P, Wong J, Glosson J, Gass J, Sun CS et al. 2007. Itraconazole IV nanosuspension enhances efficacy through altered pharmacokinetics in the rat. *International Journal of Pharmaceutics* 339: 251–260.

Rabinow BE. 2004. Nanosuspensions in drug delivery. *Nature Reviews Drug Discovery* 3: 785–796.

Rajewski RA, Traiger G, Bresnahan J, Jaberaboansari P, Stella VJ and Thompson DO. 1995. Preliminary safety evaluation of parenterally administered sulfoalkyl ether beta-cyclodextrin derivatives. *Journal of Pharmaceutical Science* 84: 927–932.

Rhodes A, Eastwood JB and Smith SA. 1993. Early acute hepatitis with parenteral amiodarone: A toxic effect of the vehicle? *Gut* 34: 565–566.

Rothschild MA, Schreiber SS and Oratz M. 1975. Effects of ethanol on protein synthesis. *Advances in Experimental Medicine and Biology* 56: 179–194.

Rowe R, Sheskey PJ and Quinn ME. 2005. *Handbook of Pharmaceutical Excipients*. New York: Pharmaceutical Press.

Rubinstein E and Lev-El A. 1980. The effect of dimethyl sulfoxide on tissue distribution of gentamicin. *Experientia* 36: 92–93.

Sakaeda T and Hirano K. 1995. O/W lipid emulsions for parenteral drug delivery II: Effect of composition on pharmacokinetics of incorporated drug. *Journal of Drug Targeting* 3: 221–230.

Salauze D and Cave D. 1995. Choice of vehicle for three-month continuous intravenous toxicology studies in the rat: 0.9% saline versus 5% glucose. *Laboratory Animals* 29: 432–437.

Shah AK and Agnihotri SA. 2011. Recent advances and novel strategies in pre-clinical formulation development: An overview. *Journal of Controlled Release* 156: 281–296.

Shekunov BY, Chattopadhyay P, Seitzinger J and Huff R. 2006. Nanoparticles of poorly water-soluble drugs prepared by supercritical fluid extraction of emulsions. *Pharmaceutical Research* 23: 196–204.

Shi Y, Porter W, Merdan T and Li LC. 2009. Recent advances in intravenous delivery of poorly water-soluble compounds. *Expert Opinion on Drug Deliver* 6: 1261–1282.

Shimomura T, Fujiwara H, Ikawa S, Kigawa J and Terakawa N. 1998. Effects of taxol on blood cells. *Lancet* 352: 541–542.

Simamora P, Pinsuwan S, Alvarez JM, Myrdal PB and Yalkowsky SH. 1995. Effect of pH on injection phlebitis. *Journal of Pharmaceutical Science* 84: 520–522

Simamora P, Pinsuwan S, Surakitbanharn Y and Yalkowsky SH. 1996. Studies in phlebitis VIII: Evaluations of pH solubilized intravenous dexverapamil formulations. *PDA Journal of Pharmaceutical Science and Technology* 50: 123–128.

Singh M and Ravin LJ. 1986. Parenteral emulsions as drug carrier system. *Journal of Parenteral Science and Technology* 40: 34–41.

Snawder JE, Benson RW, Leakey JE and Roberts DW. 1993. The effect of propylene glycol on the P450-dependent metabolism of acetaminophen and other chemicals in subcellular fractions of mouse liver. *Life Science* 52: 183–189.

Stella VJ and Rajewski RA. 1997. Cyclodextrins: Their future in drug formulation and delivery. *Pharmaceutical Research* 14: 556–567.

Stella VJ and He Q. 2008. Cyclodextrins. *Toxicologic Pathology* 36: 30–42.

Strickley RG. 2004. Solubilizing excipients in oral and injectable formulations. *Pharmaceutical Research* 21: 201–230.

Sweetana S and Akers MJ. 1996. Solubility practices for parenteral drug dosage form development. *PDA J. Pharm. Sci. Technol.* 50(5): 330–342.

Sykes E, Woodburn K, Decker D and Kessel D. 1994. Effects of Cremophor EL on distribution of taxol to serum lipoproteins. *British Journal of Cancer* 70: 401–404.

Takino T, Konishi K, Takakura Y and Hashida M. 1994. Long circulating emulsion carrier systems for highly lipophilic drugs. *Biological and Pharmaceutical Bulletin* 17: 121–125.

ten Tije AJ, Verweij J, Loos WJ and Sparreboom A. 2003. Pharmacological effects of formulation vehicles: Implications for cancer chemotherapy. *Clinical Pharmacokinetics* 42: 665–685.

Thompson DO. 1997. Cyclodextrins – enabling excipients: Their present and future use in pharmaceuticals. *Critical Reviews in Therapeutics Drug Carrier Systems* 14: 1–104.

Walters KM, Mason WD and Badr MZ. 1993. Effect of propylene glycol on the disposition of dramamine in the rabbit. *Drug Metabolism and Disposition* 21: 305–308.

Ward GH and Yalkowsky SH. 1993. Studies in phlebitis. IV: Injection rate and amiodarone-induced phlebitis. *Journal of Parenteral Science and Technology* 47: 40–43.

Woodburn K, Sykes E and Kessel D. 1995. Interactions of Solutol HS 15 and Cremophor EL with plasma lipoproteins. *International Journal of Biochemistry and Cell Biology* 27: 693–699.

Yalkowsky SH, Krzyzaniak JF and Ward GH. 1998. Formulation-related problems associated with intravenous drug delivery. *Journal of Pharmaceutical Sciences* 87: 787–796.

Yalkowsky SH. 1999. *Solubility and Solubilization in Aqueous Media*. New York: Oxford University Press.

Yasaka WJ, Sasame HA, Saul W, Maling HM and Gilette JR. 1978. Mechanisms in potentiation and inhibition of pharmacological actions of hexobarbital and zoxazolamine by glycofurol. *Biochemistry and Pharmacology* 27: 2851–2858.

Yaucher NE, Fish JT, Smith HW and Wells JA. 2003. Propylene glycol–associated renal toxicity from lorazepam infusion. *Pharmacotherapy* 23: 1094–1099.

Yokogawa K, Jin M, Furui N, Yamazak M, Yoshihara H, Nomura M, Furukawa H et al. 2004. Disposition kinetics of taxanes after intraperitoneal administration in rats and influence of surfactant vehicles. *Journal of Pharmacy and Pharmacology* 56: 629–634.

Zhang H, Yao M, Morrison RA and Chong S. 2003. Commonly used surfactant, Tween 80, improves absorption of P-glycoprotein substrate, digoxin, in rats. *Archives of Pharmaceutical Research* 26: 768–772.

chapter five

Prestudy analytical assessments
Equipment compatibility

Co-author: James Baker
AstraZeneca, UK

Introduction

It is important to assess compatibility of the chosen formulation with the dosing apparatus since any incompatibility of the formulation with these materials may lead to reduced drug delivery. The consequences would be a decrease in the therapeutic response and/or plasma kinetics (Cmax, AUC), thus adversely affecting toxicological safety evaluation. Consideration should also be given to the stability of the formulation with the vessel used to store the formulation and any filtration steps performed. Other considerations should include compatibility of the formulation with any maintenance flush, such as saline, to prevent precipitation of the formulation and serious pathological sequelae. Published literature offers many sources of information on incompatibility and stability problems that may be encountered.

Saline precipitation test

A drug can precipitate from its formulation when it is diluted with water or saline. For example, diazepam is commercially formulated in non-aqueous excipients such as propylene glycol. A dilution beyond fourfold produces a white precipitate (Murney 2008).

Some formulations are pH sensitive, and the pH may change with dilution. For example, phenytoin sodium formulation for injection, with a pH of 12, will precipitate if it is diluted in a lower pH solution, such as dextrose at pH 4.5 (Murney 2008).

For intermittent infusion programmes a saline flush is commonly used to maintain the patency of the infusion line and cannula in between infusion periods. It is, therefore, necessary to check that such saline flushes do not cause precipitation on contact with the formulation since test article residues are most likely to remain in infusion lines after the active infusion session.

The precipitation test can be conducted by making a 1:9 dilution of the highest concentration of the formulation against the maintenance flush such as saline and then shaking it. This is left to stand at room temperature for a period of time equivalent to the proposed dosing period and then examined under a bright light for signs of precipitation.

If heparinised saline is also used during the infusion or is present as part of the coating of the infusion system by the manufacturer, then an additional precipitation test is required, as some drugs, such as vancomycin, form complexes with heparin and will precipitate (Driscoll 1996).

However, if precipitation does occur it does not necessarily preclude the use of a formulation (providing it passes the blood compatibility assessment). The formulation vehicle can be used as a flush before and after administration of the formulation to prevent the contact of saline with the drug formulation, thus allowing the use of maintenance saline. This, however, does increase the work of the formulators and dosing personnel.

This work should be conducted as part of early formulation development in order to avoid problems with dosing licences from surgically prepared animals not being dosed. This additional work should be factored in when planning the study.

Stability of the formulation with the formulation storage vessel

Stability in the storage vessel is required if the formulation is to be prefilled prior to use, or syringe or bag stability to confirm that the intended concentration will be administered. Stability assessment is important, as the rate of chemical degradation doubles for every 10°C increase (the Arrhenius reaction rate theory) (Arrhenius 1889).

As formulations for vascular infusion are generally stored under refrigeration, additional stability may be required at room temperature and elevated temperatures, for example 38°C (or approximately body temperature) for implantable pumps. Another temperature effect to consider is an infusion bag being in close proximity to an animal whose body heat could cause an increase in formulation temperature. A test of stability also may be sequential and performed in succession, for example, stability to cover formulation preparation, refrigerated storage, and storage to equilibrate to room temperature and elevated temperatures.

At elevated temperatures, change in pH may also occur. This could cause the solubility of the drug in the vehicle to be reduced. At elevated temperatures ethyl vinyl acetate and latex containers exhibit an increase in the drug concentration of the formulation, believed to be due to diffusion of water from the container (Rochard et al. 1994).

Light protection also requires consideration. Visible light in a laboratory setting is around 380–780 nm (780 nm predominates), with ultraviolet wavelengths at 320–380 nm from fluorescent lighting. Planck's theory states that as the wavelength decreases, energy increases, and thus oxidation and hydrolytic degradation increase (Driscoll 1996). If sensitivity to light is a known property of the formulation then specific protection should be provided, including the use of opaque or coloured infusion lines. In cases where this is not known, protection from light should be the default.

Physical and chemical stability can be confirmed if the formulation decomposition meets the required specification for loss and is free from precipitation or colour change. In *The Handbook on Injectable Drugs*, stability is assigned based on a formulation decomposition of no more than 10% for the stated storage condition (Trissel 2003).

Some formulations may be unsuitable for storage in an infusion container such as a bag or specific infusion pump reservoir, and may need transfer just prior to infusion. Some emulsions have been noted to experience decreased stability in plastic containers compared with storage in glass (Marcus et al. 1959).

Compatibility

Compatibility assessment normally is concerned with adsorption (the drug/formulation binding to the infusion system). This can be by weak interactions, such as van der Waals forces, or by stronger ionic bonds to the materials. Adsorption occurs more strongly with drugs with an aromatic nucleus (Astier 2008). Desorption may also occur, where substances in the plastic leach into the formulation.

A compatibility trial is conducted on the same complete administration set as the one planned for use in the *in vivo* study. If not already assessed as part of the formulation development, the compatibility of filters, such as inline filters, used for the formulation preparation and administration must also be assessed, as various filter membranes are available and some may cause binding of the test item (McElnay et al. 1988). A compatibility trial is usually conducted using the lowest concentration required in the *in vivo* study (if the analytical method being used is sensitive enough to detect it). The highest concentration can also be assessed if required. The compatibility is assessed at the lowest flow rate for the duration of the study (a worst case scenario). A nominal animal weight is used for the calculation. Calibration of the infusion system to ensure that the correct flow rate is achieved may also be required prior to infusion. Temperature can also change the viscosity, resulting in a shorter or longer delivery time. If an implantable pump is used, a viscosity of up to 20 cps is recommended for the formulation (iPrecio® User Manual version 5.0.7.e).

Samples are collected at appropriate intervals and analysed usually by HPLC for loss of concentration. If a maintenance flush is used, this is also pumped through the infusion system prior to formulation priming.

It is advisable that the system be set up and operated by the person who will be performing the dosing in order to accurately reflect the procedures followed in the study.

The compatibility trial also needs to mimic the conditions that the infusion system will encounter, including light exposure and typical temperature changes.

Implantable pump compatibility can be tested using a chemical compatibility kit (e.g. Alzet®). This system allows for both the vehicle and formulation to be assessed. Spheres of the polymeric material from which the device is made are weighed and placed with a known amount of formulation for the required compatibility duration and temperatures. The vehicle and formulation vials are examined for visual changes (colour change, haziness of the solutions), and changes to the appearance of the polymeric spheres are also noted. The formulation is also analysed for any change in concentration, and the wiped spheres are reweighed to determine whether there has been a change in the weight of the spheres. This weight change could be due to adsorption or the spheres dissolving in the vehicle (www.alzet.com).

Lipid-based formulations often involve the use of strong solvents, and these can cause plastics to swell, warp, and degrade (Sastri 2010). This can require changes to the dosage.

Even if the formulation is found to react with the implanted pump, the cannula can be filled with the formulation (if this is compatible). This can then be displaced by the implantable pump reservoir filled with a compatible vehicle.

In all cases, as a general rule, compatibility is considered confirmed if the infusion concentration collected at the end of the infusion system does not differ by more than 10% from the initial formulation concentration.

Typically the following vehicles have been found to be compatible with materials used in infusion systems (amalgamation of information from iPrecio and Alzet):

- Acids with pH greater than 4.5
- Bases with pH less than 8
- Cremophor EL, up to 25% in water
- Cyclodextrins
- Dextrose, up to 5% in water or saline
- N,N-dimethylformamide (DMF), up to 25% in water
- DMSO, up to 50% in water or polyethylene glycol
- Ethanol, up to 15% in water
- Glycerol
- 1-methyl-2-pyrrolidone, up to 12.5% in water
- Phosphate buffer
- Polyethylene glycol 300 or 400, neat or in water

- Polysorbate 80, up to 2%
- Propylene glycol, neat or in water
- Saline
- Solutol, up to 30% in water
- Triacetin, up to 5% in water
- Water

To reduce incompatibility of the formulation with the materials of the infusion system, select the highest possible flow rate, smallest diameter, and shortest length of tubing.

Even when the same plastics are used, this does not guarantee that each manufacturer's plastics will perform the same with a particular drug, owing to variations stemming from the raw material manufacturer.

Choice of material

The materials most often used are silicone, polyurethane, polyethylene, polyvinylchloride, Teflon, and nylon. Flexibility, durability, chemical compatibility, biocompatibility, and thrombogenicity are the most important characteristics in selecting a material to use.

Extensive loss of test article from the infusate is most likely to occur at low flow rates and a low concentration where the formulation has a high affinity for the infusion system.

Dynamic sorption can occur where the uptake of the drug is slow and time dependent. Some drugs such as insulin and heparin react to active sites in the material. Insulin and polypeptides are strongly sorbed to glassware, polyethylene, and PVC.

Once these sites are bound, the system can be emptied and re-primed with a fresh formulation, resulting in the nominal dose being delivered. The initial uptake of the drug can be expressed as a single sorption number (log P) and be mathematically predicted with limited accuracy from the water-octanol (or hexane) partition coefficient of the formulation.

The partition coefficient is a ratio of concentrations of un-ionised compound between the two solutions. To measure the partition coefficient of ionisable solutes, the pH of the aqueous phase is adjusted such that the predominant form of the compound is un-ionised. The logarithm of the ratio of the concentrations of the un-ionised solute in the solvents is called log P. The log P value is also known as a measure of lipophilicity.

$$\text{Log } P_{oct/wat} = \text{Log}\left(\frac{[\text{solute}]_{octanol}}{[\text{solute}]_{water}^{Octanol}}\right)$$

This can aid in the selection of infusion apparatus, for example in not selecting a PVC bag (Illum and Bundgaard 1982).

A major factor in anticipating the sorption of acidic or basic drugs is the pH of the formulation. Un-ionised drug forms are the most lipophilic and the most likely sorbed to the infusion systems polymeric materials. The un-ionised drug can be controlled by pH and the pKa of the drug.

The more lipophilic the drug is, the less the sorption. Lipophilic drugs will not adsorb unless there is a pH change causing the drug become un-ionised, such as in diazepam and nitroglycerin.

Moderately hydrophobic drugs, such as midazolam, are dependent on the final pH of the formulation. At pH 3 the formulation remains as an open ring structure; at higher pH 7 the formulation becomes a closed ring and absorbs onto PVC containers (Driscoll 2002).

Sorption studies with the un-ionised drug chlormethiazole have shown that it has the greatest adsorption to PVC, followed by rubber, polyethylene, and polypropylene. Such studies have concluded that drugs with a high degree of un-ionisation, such as basic drugs with a low pKa in aqueous solutions, have the largest sorption losses in administration systems (Upton et al. 1987).

Such un-ionised drugs are more likely to be sorbed to polypropylene (typical syringe barrels). Polypropylene has good chemical resistance to most solvent and is resistant to many polar liquids. However, strong acids and oxidising agents will attack polypropylene at room temperature.

Although polypropylene syringes can be made from the same material (aside from their variations in polymer chain length), the plunger seals can vary between manufacturers. However, the formulation may bind to the other materials in the syringe (the synthetic rubber seal and silicone lubricant).

PVC is the most widely used material for clinical infusion systems. This is an amorphous plastic, which tend to be less chemically resistant than crystalline materials, as they absorb liquids more easily. Compatibility with this material has been widely published.

Drugs such as diazepam have been found to adhere to 100-mL PVC bags. This problem was solved by using 500-mL bags. This results in a reduction in the plastic surface area to volume ratio: as the volume increases, the bag area decreases.

Polypropylene and polyethylene are simple semi-crystalline plastics that do not require plasticisers, unlike PVC, and are less likely to absorb formulations. Polypropylene will swell depending on the amount of crystallisation of the polymer. High-density polyethylene has a higher crystalline structure and better chemical resistance, and it is resistant to hydrocarbon solvents (Sastri 2010).

The less polar a compound, the more rapid its permeation through polyethylene film. Work conducted by Autian has shown that functional groups attached to benzoic acid can alter permeability in polyethylene, and permeability of the compound is increased by propyl, ethyl, and methyl; whereas hydroxyl, nitro, amino, and methoxy groups decrease the permeability of the parent in the case of benzaldehyde and acetophenones (Autian et al. 1972).

PVC, on the other hand, is manufactured with plasticisers to aid flexibility. The most commonly used plasticiser for medical devices is di-2-ethylhexyl phthalate (DEHP). This plasticiser can leach into formulations containing polysorbate surfactants and ethanolic solutions, and is soluble in oils and fats, so an alternative infusion set (e.g. ethyl vinyl acetate) should be used where possible. Some products specifically state that PVC giving sets should not be used because of this leaching effect (e.g. paclitaxel injections).

Silicone and polyurethane are the most often used for cannulas. Silicone is hydrophobic, exhibits low surface tension, and is chemically and thermally stable; it is these properties that reduce the risks of coagulation of blood (Colas and Curtis 2004). It is more porous than polyurethane and not as durable; however, silicone causes less injury to the vein walls.

Thermoplastic polyurethane has no plasticisers that can be extracted. Catheters made from this material are smoother than those of silicone (Brown 1995). Further improvements in compatibility can be obtained by using blends that improve the material properties. For example, polycarbonate and polyester increase chemical resistance compared with polycarbonate only (Sastri 2010).

The FDA and EU have reported on the use of DEHP in medical devices and indicted that there may be concern about its presence in devices used with premature infants and critically ill patients. Toxicity studies in animals have demonstrated an association between prolonged exposure to DEHP and changes in hepatocellular structure and liver function. DEHP can also induce the development of hepatocellular carcinoma, and was found to be teratogenic in rats (El Abbes Faouzi et al. 1995).

References

Arrhenius S. 1889. On the reaction velocity of the inversion of cane sugar by acids. *Zeitschrift für physikalische Chemie* 4: 226.

Astier A. 2008. Compatibility of anticancer drug solutions with administering devices. *European Association of Hospital Pharmacists (EAHP)* 14.

Autian J, Nasim K and Meyer MC. 1972. Permeation of aromatic organic compounds from aqueous solutions through polyethylene. *Journal of Pharmaceutical Sciences* 61: 1775–1780.

Brown JM. 1995. Polyurethane and silicone: Myths and misconceptions. *Journal of Intravenous Nursing* 18: 120–122.

Colas A and Curtis J. 2004. Silicone biomaterials: History and chemistry and medical applications of silicones. In: *Biomaterials Science*, 2nd edition. Elsevier.

Driscoll DF. 1996. Ensuring the safety and efficacy of extemporaneously prepared infusions. *Nutrition* 12(4): 289–290.

Driscoll DF. 2002. Safety of parenteral infusions in the critical care setting. *Advanced Studies in Medicine* 2(9): 338–342.

El Abbes Faouzi M, Dine T, Luyckx M, Brunet C et al. 1995. Stability, compatibility and plasticizer extraction of miconazole injection added to infusion solutions and stored in PVC containers. *Journal of Pharmaceutical and Biomedical Analysis* 13(11): 1363–1372.

Illum L and Bundgaard H. 1981. Sorption of drugs by plastic infusion bags. *Journal of Pharmacy and Pharmacology* 33(S1): 102.

Marcus E, Kim HK and Autian J. 1959. Binding of drugs by plastics I: Interaction of bacteriostatic agents with plastic syringes. *Journal of the American Pharmaceutical Association* 48(8): 457–462.

McElnay JC, D'Arcy PF and Yahya AM. 1988. Drug sorption to glass and plastics. *Drug Metabolism and Drug Interactions* 6(1): 1–45.

Murney P. 2008. To mix or not to mix: Compatibilities of parenteral drug solutions. *Australian Prescriber* 31(4): 98–101.

Rochard E, Barthes D and Courtois P. 1994. Stability and compatibility study of carboplatin with three portable infusion pump reservoirs. *International Journal of Pharmaceutics* 101: 257–262.

Sastri, V. 2010. *Plastics in Medical Devices.* Elsevier.

Trissel LA. 2003. *Handbook on Injectable Drugs.* American Society of Health System Pharmacists.

Upton RN, Runciman WB and Mather LE. 1987. The relationship between some physicochemical properties of ionisable drugs and their sorption into medical plastics. *Australian Journal of Hospital Pharmacy* 17(4): 267–270.

Haemocompatibility

Co-author: Sophie Hill

AstraZeneca, UK

Introduction

The intravenous route of administration is frequently used in humans and animals for many classes of drugs. For any parenteral formulation of a xenobiotic, it is considered desirable that such a formulation be isotonic or at least compatible with the body fluids that are to be most immediately exposed to the formulation. Consequently, it is recommended that for solutions to be administered directly into the blood stream, tests of their effects of tonicity on blood and its constituents be performed *in vitro* prior to dosing in mammalian studies. However, such tests are not required to satisfy any particular Regulatory Guideline.

The purpose of such a *haemocompatibility study* is to assess the haemolytic potential and plasma compatibility of a test article formulation intended for intravenous administration in a given vehicle. It would be advantageous at the early stage of pharmaceutical development of a parenterally administered compound to assess the haemocompatibility of the formulation in blood from all the relevant species, including human, that might be involved in the safety assessment strategy for the product.

Objectives

A haemocompatibility study should assess various aspects of the potential for reactions between formulations and the blood. These include haemolysis, erythrocyte clumping, and plasma precipitation.

- **Haemolysis** is the disruption of erythrocytes, causing them to break open and release their contents into the surrounding fluid.
- **Erythrocyte clumping** is the clumping together of red blood cells within the plasma.
- **Plasma precipitation** is the precipitation of constituents within the plasma, giving an effect similar to clumping, which is inevitably caused by the pH of the formulation. Constituents of the test formulation can also precipitate or crystallise on contact with the blood, again due to vast differences in pH between the formulation and the blood (Yalkowsky et al. 1998).

Erythrocyte clumping and plasma precipitation tend to be more prevalent in the kidneys; both conditions can be indicated by blood being passed in the urine, the result of precipitates becoming lodged in the renal vasculature, as tubular damage can be secondary to any vascular compromise. Plasma precipitation can also occur within the dosing catheter, as when the compound is first introduced to the blood the ratio of compound to blood is much higher, consequently causing the constituents of the plasma to precipitate; this could be the result of the plasma and drug substance forming a new salt form. Severe erythrocyte clumping can cause disseminated intravascular coagulation, a very specific and severe form of coagulation in which the coagulation system is activated inside of the veins (Yalkowsky et al. 1998).

Methods of assessing haemocompatibility

With respect to how an intravenous study is conducted, the method in which a haemocompatibility test is carried out could be modified to make it more fit for purpose. The test is normally performed in a test tube, a **static method**, and therefore may not accurately reflect what happens in a dynamic *in vivo* situation where materials are being introduced directly into a flowing blood stream under pressure; this can then have an effect on a variety of variables that can invalidate the haemocompatibility study because of the differences between an *in vivo* and *in vitro* situation. Therefore, a more **dynamic method** replicating the blood stream would give a more accurate representation. These dynamic systems are a lot more difficult to set up and perform, however, and they also tend to require a larger blood volume in comparison to the static method.

Static method

The current method (Prieur et al. 1973) for a static test tube haemocompat-ability test is standardly conducted as follows for each area tested:

- **Erythrocyte clumping** – Whole blood is placed in a tube and mixed at a 1:1 ratio with test formulation; an aliquot is then placed on a glass slide and the cells are examined under low-power microscope.
- **Plasma precipitation** – Whole blood is placed in a tube and mixed at a 1:1 ratio with test formulation, and the contents are then examined for opacity with the naked eye.
- **Haemolysis** – Whole blood is placed in a tube and mixed at a 1:1 ratio with test formulation. The sample is then spun down and the haemoglobin content of the resultant supernatant is measured. The test is also performed at a 1:10 ratio of formulation to blood.

The test is based on the fact that if there is any haemolysis, red cells will release haemoglobin, which is captured in the supernatant and then measured using a HemoCue, made by Prospect Diagnostics.

Dynamic methods

Within the pharmaceutical industry a more dynamic system was developed and tested to assess haemolytic potential. Even though the dynamic test was more accurate, it was difficult to set up and required a greater volume of blood in comparison to the static method. Therefore the static method is still used in the industry today.

Dynamic human forearm method

In this system (Figure 5.1) human blood is pumped through tubing at a rate of 4 mL/min using a perfusion pump. The test vehicle is then injected into the blood flow at a rate of 0.4 mL/min, giving a vehicle-to-blood ratio of 1:10. The resulting mixture flows through a 6.6-cm length of tubing corresponding to a contact time of one second and is then quenched and analysed. The vehicles tested (unpublished in-house work) were those used in the following marketed products: Valium® injection (Roche Products), Lanoxin® injection (Glaxo Wellcome Inc.), Cordarone® X injection (Sanofi Winthrop), Vumon® injection (Bristol-Myers Squibb), and Sandimmune® injection (Sandoz Pharmaceuticals). Also tested were mixtures of PEG 400, propylene glycol, Cremaphor EL, and meglumine. The control vehicle was 0.9% w/v sodium chloride.

The infusion pumps used were the Perfusor® Compact (Braun) and the KD100® (KD Scientific); Tygon® tubing, 0.04 inch internal diameter, 50–60-mL syringes, and T connectors were used for the dynamic system.

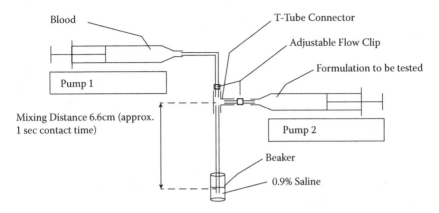

Figure 5.1 Schematic of a dynamic method of haemocompatibility assessment.

The haemoglobin concentration was measured using a HemoCue® haemoglobin analyser.

In both the static and dynamic tests, controls of 0.9% w/v sodium chloride were also assessed. The percent haemolysis was calculated by comparing the mean haemoglobin concentration of the test vehicle and the control. The pass rate for haemolysis was set at <10% in agreement with literature values (Krzyzaniak et al. 1997).

Dal Negro and Cristofori method

The dynamic test procedure is based on methods described by Prieur et al. (1973) that have been modified according to considerations outlined by Dal Negro and Cristofori (1996). The procedure involves extrapolating data obtained in an *in vitro* static model to the *in vivo* administration by taking into account the injection rate (R_{inj}) of the drug and the blood flow rate (Q_v) of the vein into which the test article is injected. An estimate of Cmax (maximum acceptable concentration at the injection site) *in vivo* can then be calculated based on the relationship between these two parameters. This relationship can be stated as follows:

$$C_{max}(mg.mL^{-1}) = \frac{R_{inj}(mg.min^{-1})}{Q_v(mL,min^{-1})}$$

R_{inj} can alternatively be expressed as mL of test solution/minute (mLsolution.minute^{-1}):

$$\frac{R_{inj}(mg.min^{-1})}{Q_v(mL,min^{-1})} = ml_{solution}.ml_{blood}^{-1}$$

Based on the data obtained in the haemolytic potential and plasma compatibility assays, the Cmax can be determined.

Blood samples.   A total of 40 mL of blood is required from each of the species under test and should be anticoagulated with EDTA. This volume should be sufficient to allow duplication of a typical three-concentration test range, three-volume test range, and three control groups (isotonic saline, vehicle, and positive control).

Test article preparation.   The stability and purity of the test article should be determined before the test is carried out to avoid hazards associated with the dilution of the formulations, the incubation temperature, and the duration of the test.

Test procedures

Haemolytic potential

Human.   For the 40-mL sample of blood from each of the volunteers, 24 tubes should be set up as follows:

Tubes 1–3 1.0 mL of whole blood + High, Intermediate, or Low volume (mL)[a] of the formulated compound (Low concentration, mg/mL)[b].

Tubes 4–6 1.0 mL of whole blood + High, Intermediate, or Low volume (mL)[a] of the formulated compound (Intermediate concentration, mg/mL)[b].

Tubes 7–9 1.0 mL of whole blood + High, Intermediate, or Low volume (mL)[a] of the formulated compound (High concentration, mg/mL)[b].

Tubes 10–12 1.0 mL of whole blood + High, Intermediate, or Low volume (mL)[a] of isotonic saline.

Tubes 13–15 1.0 mL of whole blood + High, Intermediate, or Low volume (mL)[a] 15% saponin solution in isotonic saline. (Saponin is a haemolytic agent used to lyse erythrocytes.)

Tubes 16–18 1.0 mL of whole blood + High, Intermediate, or Low volume (mL)[a] of vehicle.

Tubes 10–12 and 13–15 act as negative and positive controls, respectively, and tubes 16–18 as the vehicle controls.

[a]   The choice of volume range is based on the range of volume in mL/min expected in the programme of preclinical tests and in the clinical trials.

[b]   The concentration range should cover the lowest and highest expected to be used in the preclinical safety programme and the initial clinical trials.

Each animal species. For each pooled animal blood sample, the same ratios/concentrations as used for human blood should be investigated (ratios (v/v) of High, Intermediate and Low volume of formulated solution to blood). However, the total sample volumes are usually halved to minimise the volume of animal blood required, and each sample is prepared in duplicate as follows:

Tubes 1–6   0.5 mL of whole blood + High, Intermediate, or Low volume (mL)[a] of the formulated compound (Low concentration, mg/mL)[b].

Tubes 7–12   0.5 mL of whole blood + High, Intermediate, or Low volume (mL)[a] of the formulated compound (Intermediate concentration, mg/mL)[b].

Tubes 13–18   0.5 mL of whole blood + High, Intermediate, or Low volume (mL)[a] of the formulated compound (High concentration, mg/mL)[b].

Tubes 19–24   0.5 mL of whole blood + High, Intermediate, or Low volume (mL)[a] of isotonic saline.

Tubes 25–30   0.5 mL of whole blood + High, Intermediate, and Low volume (mL)[a] 15% saponin solution in isotonic saline. Saponin is a haemolytic agent used to lyse erythrocytes.

Tubes 31–36   0.5 mL of whole blood + High, Intermediate, or Low volume (mL)[a] of vehicle.

Tubes 19–24 and 25–30 act as negative and positive controls, respectively, and tubes 31–36 as the vehicle controls.

[a]   The choice of volume range is based on the range of volume in mL/min expected in the programme of preclinical tests and in the clinical trials.

[b]   The concentration range should cover the lowest and highest expected to be used in the preclinical safety programme and the initial clinical trials.

For the haemolytic potential assay all these tubes are incubated for 1 minute in a water bath set at 37°C, after which they are immediately centrifuged in a pre-chilled centrifuge (+4°C) at 3500 rpm (2740 g) for 5 minutes. The supernatant is removed immediately, and the amount of haemoglobin in the supernatant plasma is quantified using the cyanmethaemoglobin method (Bayer Advia 120 haematology analyser). To pass this test, the amount of haemoglobin in the test incubation should be the same as or less than the negative control, after taking into account the value of the vehicle control. A representative sample of formulated test compound is also analysed using the cyanmethaemoglobin method (Bayer Advia 120 haematology analyser), to ensure that the test compound does not interfere with the absorbance readings at the wavelength used for analysis.

Calculation of haemoglobin concentrations in samples. The value for the haemoglobin concentration in the supernatant determined by the haematology analyser needs to be corrected for the amount the blood sample was diluted by when the formulated solution or control solution was added to it. The volume change for the haemolysed red blood cells (approximately 5 µL) is considered negligible, so the entire volume change is assumed to occur only in the supernatant. This is calculated thus:

$$\text{Supernatant ratio} = 1 - \text{Haematocrit ratio}$$

Correction factor

$$= \frac{(\text{Supernatant ratio} \times \text{Blood sample volume}) \pm \text{Volume of solution added}}{(\text{Supernatant ratio} \times \text{Blood sample volume})}$$

The actual haemoglobin concentration (in g/dL) is then the measured haemoglobin concentration multiplied by the correction factor.

The relevant haematocrit values (for each species) are used for the calculations: for humans, these values are published (Kumar and Clark 2005), and for the relevant animal species they are taken from the background data in the testing facility.

By taking the ratio of dose solution to blood ($mL_{solution} \cdot mL_{blood}^{-1}$), the infusion rates to be used in the experiments ($mL.min^{-1}$) can simulate infusion rates into a human peripheral vein by multiplying by mean venous flow rate of 8 mL/min (Dal Negro and Cristofori 1996). For example:

Ratio of dose solution to blood	
($mL_{solution} \cdot mL_{blood}^{-1}$)	Infusion rate ($mL \cdot min^{-1}$)
Low concentration 0.2	1.6
Intermediate concentration 0.4	3.2
High concentration 0.8	6.4

Compatibility with plasma

For each volunteer or pooled blood sample, 1.0 mL of plasma is added to 0.8 mL of formulated compound (at 5 different concentrations) or control vehicle in a test tube.

The plasma is prepared by centrifuging the remaining volume of blood from each of the sources, under the normal conditions of 3500 rpm (2740 g), at +4°C for 5 minutes. The supernatant plasma is pipetted off into a new, clean test tube, ready for the addition of the appropriate solution as follows:

Tube 1 1.0 mL of plasma + 0.8 mL of formulated compound (Low concentration, mg/mL)[b].

Tube 2 1.0 mL of plasma + 0.8 mL of formulated compound (Intermediate concentration, mg/mL)[b].

Tube 3 1.0 mL of plasma + 0.8 mL of formulated compound (High concentration, mg/mL)[b].

Tube 4 1.0 mL of plasma + 0.8 mL of isotonic saline.

Tube 5 1.0 mL of plasma + 0.8 mL of 15% saponin solution in isotonic saline.

Tube 6 1.0 mL of plasma + 0.8 mL of isotonic phosphate buffer (pH 6.4 ± 0.3).

Tubes 4 and 5 act as a check for the compatibility of these solutions (used for the haemolytic potential experiments) with plasma.

[b] The concentration range should cover the lowest and highest expected to be used in the preclinical safety programme and the initial clinical trials.

After being incubated for 1 minute at room temperature, the tubes are visually examined for the presence of flocculation, precipitation, and coagulation.

Absence or degree of presence for each of the three categories— flocculation, precipitation, and coagulation—is recorded as follows:

0 = None present
+ = Slight
++ = Moderate
+++ = Marked

To pass this test, all tubes should be recorded 0 for each category. A potential improvement to this test could be a colourimetric assay of the tubes to provide a numerical value to the appearance of flocculation in the contents.

Discussion

The Dal Negro and Cristofori method developed in 1996 considers three different methods for assessing haemocompatibility: a dynamic method, a static method, and Prieur's method which is also static. Each has advantages and disadvantages.

Prieur's method uses a very simple 1:1 dilution in a static test tube; the blood and compound are then left to incubate for 45 minutes—a period much longer than the time the highest concentration of the test item is in contact with the blood following injection. This method does not duplicate what happens in an *in vivo* situation in that the blood and compound are left to incubate for such a long period of time at such a high concentration. This therefore highlights the need for a test that takes into account the pharmacodynamic characteristics of a formulation as well as species-specific haemodynamic parameters, such as size of animal, blood volume, and blood flow rate. But the advantage of any static method is that it is quick and easy to perform, compared with the dynamic human forearm model.

In contrast to Prieur's method, in the dynamic human forearm model, it is necessary to consider the relationship between the concentration of the test compound in the blood at the injection site, the characteristic blood flow rate for each vessel, and the injection rate of the solution. The blood flow rate in a superficial human forearm vein has been calculated to be 3–30 mL/min (Bonadonna et al. 1993a, 1993b). This is obviously a very wide range, yet it is reflective of various haemodynamic situations.

The second difference between Prieur's protocol and the one proposed by Dal Negro and Cristofori is the incubation period. In Prieur's method there is a 45-minute incubation period of one aliquot of blood and compound. As Dal Negro and Cristofori point out, in their protocol a known concentration of test compound, as it is injected, is constantly mixing with a new aliquot of new blood passing through the injected vein, as illustrated by the following equation:

$$C_{max} = R_{inj} \rightarrow C_{max}(mg/mL) = \frac{mg.min}{Q_v mL,min}$$

where the concentration of test compound in the blood at the injection site ($C_{max\ vein\ or\ artery}$ as mg/mL), the characteristic blood flow rate (Q_v as mL/min) for each vessel, and the injection rate of the solution (R_{inj} as mg/min).

The R_{inj} value can be expressed as mL of test solution/min ($mL_{solution}$/min):

$$\frac{R_{inj}(mL/min)}{Q_v(mL/min)} = mL_{solution}/mL_{blood}$$

The protocol proposed by Dal Negro and Cristofori, being a dynamic model, therefore provides only momentary contact between blood that had been pre-incubated at 37°C and the solution at Cmax.

Two experiments were set up by Dal Negro and Cristofori: one to determine the haemolytic potential of propylene glycol, and the other to determine that of two medicines on the market.

For the first experiment, 40% propylene glycol was used as the test item. The haemodynamic environment of the human forearm vein during the injection of a bolus was simulated using a model. For comparison, the test was then repeated statically using the same volumetric ratios, and then repeated again using Prieur's protocol. The results showed that there was potassium leakage of approximately 6% and 9% in the dynamic and static tests, respectively, whereas in Prieur's method this was approximately 60%. The haemolysis test results showed a similar pattern, with the dynamic and static methods producing a result of approximately 2%, and Prieur's with a result of 41%.

The results of these tests, with the dynamic and static models rendering similar results in both the potassium and haemolysis tests whereas Prieur's model produced significantly higher results, indicate that incubation time has a dramatic effect on the amount of extracellular Hgb and K. Dal Negro and Cristofori then decided to quench the process after 30 seconds. The experiment included quench times of 15s, 30s, 1 min and 45 min and they came to the conclusion that 30s was an adequate time to quench as it allowed all operations to be completed and is more realistically similar to an *in vivo* situation.

The second experiment involved determining the haemolytic potential of two medicines that were already on the market, Valium and Lanoxin. These were chosen because of the vehicles' being 40% propylene glycol, thereby minimising variability in the experiment. The experiments were repeated as in the first experiment. The instructions for the intravenous administration state that they should be administered slowly (<1 mL/min).

The results showed that there were again no noticeable differences between the dynamic and static tests, and for the control (propylene glycol in 5% dextrose) and Lanoxin even Prieur's model gave a similar result. Valium gave a far higher haemolysis and potassium leakage result in Prieur's method. No noticeable differences were obtained with any of the three tests; the vehicle does not induce haemolysis or K leakage, and Valium gave similar results in the static and dynamic tests for both haemolysis and K leakage, except in Prieur's model, where the results were significantly higher. The Lanoxin gave similar results in all three tests for haemolysis, whereas the K leakage test produced a higher result in Prieur's model.

Conclusion

The result of a haemocompatibility study is dependent on the structure/class of the compound and also the constitution of the formulation being tested. Formulations play a key role in assessing the biological properties of a molecule during drug discovery. Maximising exposure is the primary objective in early animal experimentation, so that the pharmacokinetics, pharmacodynamics, and toxicological signals can be put into context with the biological response to the specific targets. Drug discovery is a very complex process, and pharmaceutical organisations typically adopt a stepwise filtering approach to systemically select compounds that have desirable properties for the appropriate therapy area. For intravenous formulations, the dose volume used, stability of the formulation, pH, viscosity, osmolality, buffering capacity, and biocompatibility of formulations are all factors to consider.

As stated before, Prieur's model, with its 45-minute incubation, is not an accurate reflection of what happens *in vivo* during the intravenous infusion of a xenobiotic. The dynamic methods used by both Dal Negro and Cristofori (1996) and the dynamic human forearm method shown in Figure 5.1 earlier in this chapter are potentially more accurate than the static method, but they are more difficult to set up and it is difficult to predict an actual contact time of the formulation to blood.

All this highlights the need for careful study design, as Dal Negro and Cristofori's two experiments, discussed earlier in this chapter, both produced haemolytic results, showing that an experiment is needed in which the potential of a test solution to affect red blood cell integrity can be evaluated, and eventually modulated, by changing the injection rate value, and as a consequence, the Cmax.

Prieur's method, because it is a fixed blood to formulation ratio, often reproduces situations in which the drug concentration is far higher than that which occurs *in vivo*. False positives are, therefore, difficult to explain, as none of the parameters involved in an intravenous injection is taken into account. In contrast, according to the results obtained, Dal Negro and Cristofori's static method is considered to be an adequate and reliable representation of an *in vivo* situation.

Consequently, although the results obtained from the dynamic and static tests are essentially comparable, it is considered that the dynamic test approach, per the Dal Negro and Cristofori (1996) procedure, is the more accurate and preferred approach to follow. This particular test, as discussed, is more difficult to set up because of the estimations that have to be made with regard to formulation:blood contact time and concentration variation. However, to avoid the consequences of haemo-incompatibility upon first dose to animals, it is in the best interest of animal welfare to ensure that haemocompatibility of test formulations is examined in detail before undertaking vascular infusion studies.

References

Bonadonna RC, Saccomani MP, Seely L, Starick-Zych K, Ferrannini E, Cobelli C and DeFronzo RA. 1993. Glucose transport in human skeletal muscle: The *in vivo* response to insulin. *Diabetes* 42: 191–198.

Bonadonna RC, Saccomani MP, Cobelli C and DeFronzo RA. 1993. Effect of insulin on system A amino acid transport in human skeletal muscle. *J. Clin. Invest.* 91: 514–521.

Dal Negro G and Cristofori P. 1996. A new approach for evaluation of the *in vitro* haemolytic potential of a solution of a new medicine. *Comp. Haematol. Int.* 6: 35–41.

Dirckx JH, ed. 2001. Appendix 118 (Laboratory reference range values), in: *Stedman's Concise Medical Dictionary for the Health Professions*, 4th edition. Lippincott, Williams and Wilkins.

Kumar P and Clark ML. 2005. *Clinical Medicine*, 5th edition. Elsevier.

Krzyzaniak J, Alvarez Nunez FA, Raymond DM et al. 1997. Lysis of human red blood cells. 4. Comparison of *in vitro* and *in vivo* hemolysis data. *J. Pharm. Sci.* 86 (11): 1215–1217.

Prieur DJ, Young DM, Davis D et al. 1973. Procedures for preclinical toxicologic evaluation of cancer. *Cancer Chemother. Rep.* 4: 1–39.

Yalkowsky SH, Krzyzanick J and Ward GH. 1998. Formulation-related problems associated with intravenous drug delivery. *J. Pharm. Sci.* 87: 787–796.

Annex: Common excipients and vehicles

Introduction

This Annex contains information on a number of commonly used vehicles and excipients in intravenous infusion delivery formulations. The information given will relate to the physico-chemical parameters, at various concentrations, discussed in previous chapters, along with some toxicological considerations to be mindful of when conducting non-clinical safety assessment studies.

The data are set out in the form of Test Substance Data Sheets in such a way that the specific chemical can be readily identified in all references.

Vehicle/Excipient Data Sheet

Chemical Name	**Acetic Acid, glacial (Glacial Acetic Acid)**
CAS ID Number	64-19-7
Molecular Weight	60.05
Molecular Formula	$C_2H_4O_2$
Solubility in Water	Miscible with water at 20°C
pH	3.4 (0.01 M aqueous solution)
Osmolality	Normally hypertonic to plasma/serum at approximately 713 ± 6 mOsm.KgH$_2$O^{-1} dependent on sodium content
Surface Tension	25.5–29.0 mN/m at 20°C
Viscosity	1.22 mPa.s at 20°C
Stability	Acetic acid should be stored in an airtight container in a cool, dry place. (From *Handbook of Pharmaceutical Excipients*)
Toxicity	Acetic acid is widely used in pharmaceutical applications primarily to adjust the pH of formulations and is thus generally regarded as relatively non-toxic and non-irritant. However, glacial acetic acid or solutions containing over 50% w/w acetic acid in water or organic solvents are considered corrosive and can cause damage to skin, eyes, nose, and mouth. LD$_{50}$ (Mouse IV): 0.525 g/kg LD$_{50}$ (Rabbit dermal): 1.06 g/kg LD$_{50}$ (Rat Dral): 3.31 g/kg (From *Handbook of Pharmaceutical Excipients*)
Other Comments	Acetic acid is widely used as an acidifying agent. (From *Handbook of Pharmaceutical Excipients*)

Vehicle/Excipient Data Sheet

Chemical Name	**Dextrose (Glucose, Glucose Monohydrate)**
CAS ID Number	50-99-7 (anhydrous); 5996-10-1 (monohydrate)
Molecular Weight	180.16 (anhydrous); 198.17 (monohydrate)
Molecular Formula	$C_6H_{12}O$ (anhydrous); $C_6H_{12}O_6 \cdot H_2O$ (monohydrate)
Solubility in Water	Very soluble – 1 part in less than 1 part water at 20°C
pH	5.9 (10% w/v aqueous solution of anhydrous dextrose)
Osmolality	5.05% w/v anhydrous dextrose is iso-osmotic with serum
Surface Tension	42 mN/m at 20°C
Viscosity	5% solution 1.0385 mPa.s at 20°C
Stability	Dextrose has good stability under dry storage conditions. Aqueous solutions may be sterilised by autoclaving; however, excessive heating can cause a reduction in pH and caramelisation of solutions. (From *Handbook of Pharmaceutical Excipients*)
Toxicity	LD_{50} (Rat Oral): 25.8 g/kg LD_{50} (Mouse IV): 9.0 g/kg LD_{50} (Rabbit IV): 35.0 g/kg
Other Comments	Dextrose is widely used in parenteral solutions to adjust tonicity or where it may function both as a volume expander and a means of parenteral nutrition. Care must be taken when incorporating into chronic intravenous delivery formulations as it has high potential for the development of bacterial colonies within the formulation (Salauze and Cave 1995).
	Dextrose solutions are incompatible with a number of pharmaceuticals such as cyanocobalamin, kanamycin sulphate, novobiocin sodium, and warfarin sodium. Erythromycin gluceptate is unstable in dextrose solutions at a pH less than 5.05. Decomposition of B-complex vitamins may occur if they are warmed with dextrose. In the aldehyde form, dextrose can react with amines, amides, amino acids, peptides, and proteins. Brown coloration and decomposition occur with strong alkalis. (From *Handbook of Pharmaceutical Excipients*)

Vehicle/Excipient Data Sheet	
Chemical Name	**Dimethyl Sulfoxide (DMSO)**
CAS ID Number	67-68-5
Molecular Weight	78.13
Molecular Formula	C_2H_6OS
Solubility in Water	Miscible with water with evolution of heat at 20°C; also miscible with ethanol (95%), ether, and most organic solvents; immiscible with paraffins, hydrocarbons. Practically insoluble in acetone, chloroform, ethanol (96%), and ether.
pH	8.5 (for a 50:50 mixture with water)
Osmolality	2% is iso-osmotic.
Surface Tension	53.6 mN/m at 20°C
Viscosity	1.996 mPa s at 20°C
Stability	Reasonably stable to heat, but upon prolonged reflux it decomposes slightly to methyl mercaptan and bismethylthiomethane. This decomposition is aided by acids and is retarded by many bases. When heated to decomposition, toxic fumes are emitted.
	At temperatures between 40 and 60°C, it has been reported that DMSO suffers a partial breakdown, which is indicated by changes in physical properties such as refractive index, density, and viscosity.
	DMSO should be stored in airtight, light-resistant glass containers, and contact with plastics should be avoided. (From *Handbook of Pharmaceutical Excipients*)
Toxicity	DMSO is a highly polar substance that has exceptional solvent properties for both organic and inorganic chemicals and is widely used as an industrial solvent. It is a scavenger of hydroxyl radicals and is reported to have a wide spectrum of pharmacological activity, including
	Membrane penetration (Lawrence and Goodnight 1983; Olver et al. 1988; Wood and Wood 1975)
	Anti-inflammatory effects (Trice and Pinals 1985; Haschek et al. 1989)
	Local analgesia
	Intracranial hypertension (Marshall et al. 1984; Karaca et al. 1991)
	Weak bacteriostasis
	Diuresis
	Vasodilation
	Dissolution
	Free radical scavenging

(*Continued*)

The principal use of DMSO is as a vehicle for drugs; it aids penetration of the drug into the skin and may also enhance the drug's effect. It is also used as a 50% aqueous solution for bladder instillation for the symptomatic relief of interstitial cystitis (Fowler 1981; Perez-Marrero et al. 1988; Ryan and Wallace 1990). DMSO has been administered orally, intravenously, and topically for a wide range of indications including cutaneous and musculoskeletal disorders, but evidence of beneficial effects is limited.

Acute toxicology data:

LD_{50} (Dog IV): 2.5 g/kg

LD_{50} (Rat IP): 8.2 g/kg

LD_{50} (Rat IV): 5.3 g/kg

LD_{50} (Rat Oral): 14.5 g/kg

LD_{50} (Rat SC): 12.0 g/kg

LD_{50} (Mouse IP): 2.5 g/kg

Chronic toxicology data:

As a consequence of its widespread activity, DMSO also has a wide range of reported side effects or toxicity (Brobyn 1975; Willhite and Katz 1984; Brayton 1986). High concentrations of DMSO applied to the skin may cause burning discomfort, itching, erythema, vesiculation, and urticaria. Systemic effects, which may occur after administration by any route, include gastrointestinal disturbances, drowsiness, headache, and hypersensitivity reactions. Intravascular haemolysis has been reported following intravenous administration (Muther and Bennett 1980). Local discomfort and spasm may occur when given by bladder instillation. When used as a penetrating basis for other drugs, DMSO may enhance their toxic effects (Kocsis et al. 1975; Rayburn et al. 1986).

DMSO administered by intravenous infusion to patients caused transient haemolysis and haemoglobinuria. Infusion strengths greater than 10% were associated with grossly discoloured urine but there was no evidence of kidney damage (Muther and Bennett 1980).

Other side effects reported in patients include raised liver and muscle enzyme concentrations, mild jaundice, acute renal tubular necrosis, deterioration in level of consciousness, and incidence of cerebral infarction (Yellowlees et al. 1980). Acute, reversible neurological deterioration has also been reported in a patient following intravenous DMSO (Bond et al. 1989). In animal studies, toxicities following long-term oral administration of a 50% aqueous solution to rats and dogs, and dermal

(Continued)

application of a 50% aqueous solution to rabbits and pigs resulted in minor disturbances in bodyweight and haematological parameters. However, there was reported a physiological diuretic response and, more importantly, a specific effect on the eye and, in particular, the lenticular nucleus (Noel et al 1975). Haemolytic and precipitation potential tests have been carried out to assess blood compatibility of a range of solvents, including DMSO, in various inbred mouse strains (Montaguti et al. 1994). Dilution in water is suggested to ameliorate blood compatibility, and suggested doses are given.

Raised serum enzyme levels have been reported in rats following a single intraperitoneal injection (Yoshikawa 1985).

Other Comments   None

Vehicle/Excipient Data Sheet

Chemical Name	**Glycerin (Glycerol, Propane 1,2,3 triol)**
CAS ID Number	56-81-5
Molecular Weight	92.09
Molecular Formula	$C_3H_8O_3$
Solubility in Water	Miscible in water at 20°C
pH	pH 3.97 in distilled form
Osmolality	2.6% v/v solution is iso-osmotic with serum
Surface Tension	63.4 mN/m (63.4 dynes/cm) at 20°C
Viscosity	5% v/v solution in water 1.143 mPa.s at 20°C
Stability	Glycerin is hygroscopic. Pure glycerin is not prone to oxidation by the atmosphere under ordinary storage conditions, but it decomposes on heating with the evolution of toxic acrolein. Mixtures of glycerin with water, ethanol (95%), and propylene glycol are chemically stable.
	Glycerin should be stored in an airtight container, in a cool, dry place.
	(From *Handbook of Pharmaceutical Excipients*)
Toxicity	Glycerin occurs naturally in animal and vegetable fats and oils that are consumed as part of a normal diet.
	LD_{50} (Guinea pig Oral): 7.8 g/kg
	LD_{50} (Mouse IP): 8.7 g/kg
	LD_{50} (Mouse IV): 4.3 g/kg
	LD_{50} (Mouse Oral): 4.1 g/kg
	LD_{50} (Mouse SC): 0.1 g/kg
	LD_{50} (Rabbit IV): 0.05 g/kg
	LD_{50} (Rabbit Oral): 27.0 g/kg
	LD_{50} (Rat IP): 4.4 g/kg
	LD_{50} (Rat IV): 5.6 g/kg
	LD_{50} (Rat Oral): 12.6 g/kg
	LD_{50} (Rat SC): 0.1 g/kg
	(From *Handbook of Pharmaceutical Excipients*)
Other Comments	None

Vehicle/Excipient Data Sheet

Chemical Name	**Glycofurol**
CAS ID Number	31692-85-0
Molecular Weight	190.24
Molecular Formula	$C_9H_{18}O_4$
Solubility in Water	Miscible with water at 20°C
pH	No data available
Osmolality	50% solution has an osmolality of 380 mOsm/kg
Surface Tension	40.1 mN/m at 25°C
Viscosity	8–18 mPa.s at 20°C
Stability	Stable if stored under nitrogen in a well-closed container protected from light, in a cool, dry place. (From *Handbook of Pharmaceutical Excipients*)
Toxicity	Glycofurol is mainly used as a solvent in parenteral pharmaceutical formulations and is generally regarded as a relatively non-toxic and non-irritant material at the levels used as a pharmaceutical excipient. Glycofurol can be an irritant when used undiluted; its tolerability is approximately the same as propylene glycol.
	Glycofurol may have an effect on liver function and may have a low potential for interaction with hepatotoxins or those materials undergoing extensive hepatic metabolism. LD_{50} (Mouse IV): 3.5mL/kg
Other Comments	Glycofurol is used as a solvent in parenteral products for intravenous of intramuscular injection in concentrations up to 50% v/v. (From *Handbook of Pharmaceutical Excipients*)

Vehicle/Excipient Data Sheet

Chemical Name	**Hydrochloric Acid**
CAS ID Number	7647-01-0
Molecular Weight	36.46
Molecular Formula	HCl
Solubility in Water	Miscible with water at 20°C
pH	0.1 (10% v/v aqueous solution)
Osmolality	No data available
Surface Tension	72 mN/m as a 10% v/v aqueous solution at 20°C
Viscosity	1.19 mPa.s as a 10% aqueous solution at 20°C
Stability	Hydrochloric acid should be stored in a well closed, glass, or other inert container at a temperature below 30°C. Storage in close proximity to concentrated alkalis, metals, and cyanides should be avoided. (From *Handbook of Pharmaceutical Excipients*)
Toxicity	LD_{50} (Mouse IP): 1.4 g/kg LD_{50} (Rabbit Oral): 0.9 g/kg
Other Comments	Hydrochloric acid is widely used as an acidifying agent. (From *Handbook of Pharmaceutical Excipients*)

Vehicle/Excipient Data Sheet

Chemical name	**Hydroxypropyl-beta-cyclodextrin**
CAS ID Number	128446-35-5
Molecular Weight	1135
Molecular Formula	$C_{42}H_{70}O_{35}(C_3H_6O)_x$ (where x = 7 molar substitution)
Solubility in Water	Greater than 1 in 2 parts of water at 25°C
	1.85 g/100mL at 25°C (Loftsson and Brewster 1996)
pH	5–8 in a 20 g/L solution at 20°C for Cavasol W7 HP
	Pharma (may vary between manufacturers)
Osmolality	28% w/v is typically iso-osmotic with serum.
Surface Tension	No data available
Viscosity	Not viscous at 20°C at concentrations up to 30%
Stability	Store in well-closed containers.
	(From *Handbook of Pharmaceutical Excipients*)
Toxicity	Intravenous administration studies have been performed in rats, mice, dogs, and monkeys at various dose levels after single or repeated doses for up to 90 days. In intravenous infusion studies in rats for 24 hours/day for 7 days, doses of 225 mg/kg/day of 11.25% w/v or 2400 mg/kg/day of 5% w/v HP-β-CD, vacuolation of the renal proximal convoluted tubule was reported along with foamy alveolar macrophage infiltration in the lung. However, in 14-day and 90-day studies, dosing on alternate days with 200 mg/kg of 20% w/v HP-β-CD to rats and *Cynomolgus* monkeys, no reactions to the treatment were observed (Brewster et al. 1990). Two 90-day studies by intravenous dosing were performed in rats and dogs at doses of 50, 100, or 400 mg/kg/day. In the rat, there were no adverse findings at 50 mg/kg/day. At 100 mg/kg/day, there were minimal histological changes in the urinary bladder (swollen epithelial cells), swollen and granular kidney tubular cells, and an increase in Kupffer cells in the liver. At 400 mg/kg/day, there was a decrease in body weight and food consumption, increase in water consumption, a decrease in haematocrit, haemoglobin, and erythrocytes, and an increase in creatinine, bilirubin and aspartate, and alanine aminotransferase plasma levels (AST and ALT, respectively). The weight of the spleen, adrenals and kidneys were increased and histopathological changes included foamy cells in the lungs, increased spleen red

(Continued)

pulp hyperplasia, increased rough endoplasmic RES aggregates in the liver, and increased round cells in the Kieran space of the liver. Following a one-month recovery period, most of the toxicological changes had reversed. However, there were still small elevations in ALT and AST levels and only a partial reversal of the urinary tract and lung changes. In the dog, there were no adverse effects at 50 or 100 mg/kg/day. At 400 mg/kg/day, there were slight increases in ALT and AST and total bilirubin. Histological changes were seen in the lung (foamy cell), and there were swollen epithelial cells of the urinary bladder and renal pelvis. At the end of the recovery period, the toxicological changes had completely reversed except for an incomplete recovery of the swollen renal pelvis epithelium (Coussement et al. 1990).

Other Comments   Hydroxypropyl-beta-cyclodextrin is widely used parenterally as a solubility enhancer/complexing agent generally at concentration <45% w/v.

Vehicle/Excipient Data Sheet

Chemical Name	**Lecithin**
CAS ID Number	8002-43-5
Molecular Weight	The USP32-NF27 describes lecithin as a complex mixture of acetone-insoluble phosphatides that consists chiefly of phosphatidylcholine, phosphatidylethanolamine, phosphatidylserine, and phosphatidylinositol, combined with various amounts of other substances such as triglycerides, fatty acids, and carbohydrates as separated from a crude vegetable oil source.
	The composition of lecithin (and hence also its physical properties) varies enormously depending upon the source of the lecithin and the degree of purification. Egg lecithin, for example, contains 69% phosphatidylcholine, and 24% phosphatidylethanolamine, while soybean lecithin contains 21% phosphatidylcholine, 22% phosphatidylethanolamine, and 19% phosphatidylinositol, along with other components. *(From Handbook of Pharmaceutical Excipients)*
Molecular Formula	See above
Solubility in Water	Lecithins hydrate to form emulsions in water at 20°C.
pH	Most effective at pH 4.0
Osmolality	383 mOsmol/kg at pH 4.5 and 20°C
Surface Tension	Lecithin has the effect of reducing surface tension between particles in oil-in-water emulsions.
Viscosity	100 stokes maximum at 25°C
Stability	Lecithins decompose at extreme pH. They are also hygroscopic and subject to microbial degradation. When heated, lecithins oxidise, darken, and decompose. Temperatures of 160–180°C will cause degradation within 24 hours.
	Fluid or waxy lecithin grades should be stored at room temperature or above; temperatures below 10°C may cause separation.
	All lecithin grades should be stored in well-closed containers protected from light and oxidation. Purified solid lecithins should be stored in tightly closed containers at subfreezing temperatures. *(From Handbook of Pharmaceutical Excipients)*
Toxicity	Lecithin is a component of cell membranes and is, therefore, consumed as a normal part of the diet. *(From Handbook of Pharmaceutical Excipients)*
Other Comments	Lecithins are mainly used in pharmaceutical products as dispersing, emulsifying, and stabilising agents, and are included in intramuscular and intravenous injections. *(From Handbook of Pharmaceutical Excipients)*

Vehicle/Excipient Data Sheet	
Chemical Name	**Meglumine (1-Deoxy-1-(methylamino)-d-glucitol, N-Methylglucamine)**
CAS ID Number	6284-40-8
Molecular Weight	195.21
Molecular Formula	$C_7H_{17}NO_5$
Solubility in Water @ 20°C	1 in 1
pH	10.5 (1% w/v aqueous solution)
Osmolality	5.02% w/v aqueous solution is iso-osmotic with serum.
Surface Tension	No data available
Viscosity	6.17 mPa.s at 60% w/v at 25°C
Stability	Meglumine does not polymerise or dehydrate unless heated above 150°C for prolonged periods. The bulk material should be stored in a well-closed container in a cool, dry place. Meglumine should not be stored in aluminium containers since it reacts to evolve hydrogen gas; it discolours if stored in containers made from copper or copper alloys. Stainless steel containers are recommended. (From *Handbook of Pharmaceutical Excipients*)
Toxicity	LD_{50} (Mouse IP): 1.68 g/kg Known to be irritant to skin and eyes, but very little data available (From *Handbook of Pharmaceutical Excipients*)
Other Comments	Meglumine is widely used in parenteral formulations and is generally regarded as a non-toxic material at the levels usually employed as an excipient. (From *Handbook of Pharmaceutical Excipients*)

Vehicle/Excipient Data Sheet	
Chemical Name	**Poloxamers (α-Hydro-ω-hydroxypoly(oxyethylene) poly(oxypropylene)poly(oxyethylene) block copolymer)**
CAS ID Number	9003-11-6
Molecular Weight	The poloxamer polyols are a series of closely related block copolymers of ethylene oxide and propylene oxide conforming to the general formula $HO(C_2H_4O)_a(C_3H_6O)_b(C_2H_4O)_aH$
Molecular Formula	See above
Solubility in Water @ 20°C	Poloxamer 188 is freely soluble in water.
pH	5.0–7.4 for a 2.5% w/v aqueous solution
Osmolality	2.6% v/v solution is iso-osmotic with serum.
Surface Tension	63.4 mN/m (63.4 dynes/cm) at 20°C
Viscosity	5% v/v solution in water 1.143 mPa.s at 20°C
Stability	Poloxamers are stable materials. Aqueous solutions are stable in the presence of acids, alkalis, and metal ions. However, aqueous solution supports mould growth. The bulk material should be stored in a well-closed container in a cool, dry place. (From *Handbook of Pharmaceutical Excipients*)
Toxicity	Poloxamers are used in a variety of oral, parenteral, and topical pharmaceutical formulations, and are generally regarded as non-toxic and non-irritant materials. Poloxamers are not metabolised in the body. Following intravenous administration in rats and dogs, P188 has an elimination half-life of about 3–6 hours. P188 is eliminated in the urine and is not metabolised (Kibbe 2000; BASF 1978). Not all poloxamers are best suited to intravenous formulations. For example, the low toxicity and chemical nature of P407 prompted the use of the product in the development of a sustained release formulation of recombinant human growth hormone (rhGH). This formulation was evaluated in a rat model (Katakam et al. 1997). Since P407 undergoes reverse thermal gelation, it can be injected as a solution but then it forms a gel matrix at body temperature.

(Continued)

Poloxamer 188 (Pluronic F68)

Pharmacology

P188 was used experimentally as long ago as 1968 by Hymes et al. to protect blood from trauma and to enhance circulation. In 1971, Hymes et al. using a model of **haemorrhagic shock** in groups of 24–27 anaesthetised male beagle dogs, compared the infusion of a 0.4 g% concentration of P188 dissolved in Ringer's lactate together with the shed blood with Ringer's lactate and shed blood alone. There was significantly higher survival for animals treated with the P188 formulation, and these animals produced significant amounts of urine also. The blood surface tension was markedly reduced in the P188-treated animals, which was suggested to have an important role in restoring the microcirculation in this experimental group.

Platelet aggregometer studies indicate that P188 decreases **platelet aggregation** and enhances deaggregation (Nalbandian et al. 1969). By the use of microcinephotography, it has been demonstrated that P188 prevents aggregation of microthrombi of platelets and, if the pluronic is infused early after thrombosis, it deaggregates platelets and restores the obstructed microcirculation. It could be postulated that the use of P188 re-established a portion of the microcirculation and counteracted, in part, the effect of disseminated intravascular coagulation in haemorrhagic shock. Large amounts of P188 in the perfusate for open-heart surgery and experimental cardiopulmonary bypass has not been associated with post-operative haemorrhage (Hymes et al. 1968). In a subsequent clinical study (Grover et al. 1969) groups of patients undergoing open-heart surgery or cardiopulmonary bypass were given injections of 0.6 mg/mL blood volume of P188, which had the effect of significantly decreasing **blood viscosity** and **platelet adhesiveness**. *In vitro* studies on the effects of P188 on surface tension and viscosity of blood and plasma (Manja et al. 1974) in blood samples from healthy human donors implies that a 0.4 g% P188 solution reduces the surface tension of plasma, perhaps by the weakening of cohesive forces and, therefore, of whole blood. The relative viscosity of whole blood was significantly

(Continued)

reduced, but that of plasma remained unaltered at this concentration. Further blood viscosity studies were performed in groups of between 10 and 20 mongrel dogs (Grover et al. 1973) in which animals were given P188 (0.6 mg/mL perfusion volume of a 10% solution) prior to, during, or after one hour of total cardiopulmonary bypass. In all groups, administration of P188 was found to lower blood viscosity without affecting haemodilution. However, the combination of haemodilution coupled with repeated hourly doses of P188 caused the greatest and statistically the most significant reduction in viscosity.

As a result of studies in rats, these changes in haemodynamics caused by P188 are not apparently associated with any **haemolytic properties**. Incubation of blood from mice, rats, rabbits, and hamsters with up to 4.0% (w/v) of commercial or purified P188 produced no detectable haemolysis (Lowe et al. 1995). Haemolysis did occur with concentrations of commercial P188 above 4.0% (w/v). This was maximal with rat blood incubated with 10.0% (w/v) P188, where the mean haemolysis was $4.7 \pm 1.5\%$; the mean haemolysis in rat blood was reduced to $0.5 \pm 0.3\%$ following incubation with the purified fraction.

Utilising the haemorrhagic shock model in 16 adult mongrel dogs (8 control – 0.9% NaCl; and 8 treated with 5% P188 at 1 ml/kg), the effects on central circulatory dynamics were investigated (Grover et al. 1974). It was found that P188 did not significantly alter cardiac output, mean arterial pressure, central venous pressure, pulmonary artery pressure, pulmonary wedge pressure, heart rate, total peripheral resistance, pulmonary vascular resistance, stroke volume, left ventricular stroke work, left ventricular work, or carotid artery flow when administered without volume replacement in experimental haemorrhagic shock. However, P188 did significantly increase renal artery flow. Therefore, this change, together with increases in mesentery microcirculation and urinary output reported in other experiments (Nalbandian et al. 1969), would appear not to be related to a central ionotropic effect but rather to the peripheral rheological properties of P188.

(Continued)

Interestingly, an *in vitro* study measuring plasma and whole blood viscocity of RheothRx®, an intravenous formulation of P188, in blood samples from healthy human subjects is in disagreement with these findings in animal studies (Lechmann and Reinhart 1998). RheothRx solution contains 150 mg P188 per mL and, in this experiment, plasma was incubated with 0, 0.75, 3.75, or 18.75 mg/mL RheothRx and whole blood with a constant haematocrit of 41.4% with 0, 0.4, 2, and 10 mg/mL RheothRx at 37°C. In contrast to other studies, no favourable effect of RheothRx on plasma and whole blood viscosity was found. At the highest RheothRx concentration, an actual increase in high and low shear viscosity was observed, and erythrocyte morphology remained unchanged. These data, although on normal blood *in vitro*, suggest that the positive effects of P188 *in vivo* may not be caused by improved flow properties of blood, but could rely on other mechanisms, for example an interaction with thrombolytic substances (Hunter et al. 1990), platelet aggregation (Benner et al. 1973), and/or activation of neutrophils (Babbitt et al. 1990; Bajaj et al. 1989). However, in a clinical study in two groups of patients with acute myocardial infarction and who were not eligible for thrombolytic therapy (Manyard et al. 1998), RheothRx treatment did not decrease infarct size or formally alter the outcome. In fact, increased frequencies of renal and left ventricular dysfunction were observed, which were not explained.

With regards to the effects on **coagulation**, P188 has been extensively studied in a variety of test systems (see Toxicology section) to investigate potential anticoagulant activity. P188 has no anticoagulant effects at concentrations up to 10–20 times greater than the anticipated therapeutic range on prothrombin time (PT), partial thromboplastin time (PTT), activated partial thromboplastin time (aPTT), or thrombin clotting time in either human or animal plasma. Despite this, inhibition of coronary artery thrombosis was reported in pigs infused with P188 (Robinson et al. 1990). Twenty-two normal juvenile pigs received a bolus injection of heparin (100 U/kg)

(Continued)

and wire coil stents in the left anterior descending coronary artery, and were randomised to infusion of P188 (RheothRx bolus of 50 mg/kg followed by infusion of 25 mg/kg/hour) or equivalent volume of 0.9% NaCl. P188 significantly inhibited thrombosis as determined by morphometry of arterial specimens. P188 infusion did not affect bleeding time or platelet aggregation but did reduce blood viscosity. Dosing studies in 13 additional pigs that received 0.9% NaCl as control or a single arterial dose of 50, 100, or 200 mg/kg showed a modest drop in white blood cell count with P188 injection, which may reflect intravascular leukocyte margination. In contrast to some human studies with P188 (Vercellotti et al. 1982) or emulsion containing P188, a drop in neutrophil count attributable to activation of complement was not observed. However, in studies of a dog model of acute myocardial infarction (AMI), alterations in neutrophil function were induced by P188 (Schaer et al. 1994). P188 concentrations of 0.5 to 2.0 mg/mL produced significant inhibition of neutrophil chemotaxis, which could result in decreased neutrophil infiltration into the postischaemic myocardium. In fact, Justicz et al. (1991) demonstrated in their canine model of myocardial infarction that treatment with P188 was associated with a significant reduction in neutrophil infiltration 24 hours after reperfusion. The mechanism by which P188 inhibits neutrophil chemotaxis may result from P188-induced neutrophil activation, concomitant release of superoxide anion, and subsequent deactivation (Forman et al. 1992). P188 might also reduce neutrophil-mediated injury by interfering with neutrophil adhesion.

P188 has been studied in several canine models of **myocardial ischaemia** and reperfusion (Bajaj et al. 1989; Forman et al. 1992; Schaer et al. 1990; Tokioka et al. 1987; Simpson et al. 1988). In these studies, P188 shortened the time to reperfusion and lengthened the time to reocclusion by 40–60% compared to controls. Following thrombolysis, P188-treated animals demonstrated significantly increased blood flow in both the subepicardial and subendocardial ischaemic zones 120 minutes after thrombolysis, compared to control animals.

(Continued)

In a closed-chest model (Schaer et al. 1994) three groups of adult male mongrel dogs were prepared. One group of 13 dogs received an intravenous bolus injection of P188 at 75 mg/kg 15 minutes before reperfusion. This was immediately followed by a continuous intravenous infusion of P188 at 150 mg/kg/hour for 4 hours (rate of 1.07 ml/kg/hour) after which the infusion was changed to half normal sterile saline (4.5 mg/ml) given at the same rate for the subsequent 44 hours. The infusion was then discontinued for the final 24 hours of reperfusion. A second group of 13 dogs received the same bolus injection of P188 as above 15 minutes before reperfusion followed by a continuous intravenous infusion at 150 mg/kg/hour for 48 hours at the same rate as above. The infusion was discontinued for the final 24 hours of reperfusion. The third group of 12 dogs received a sterile saline solution (4.5 mg/mL) at 1.07 ml/kg/hour for 48 hours and acted as controls. A 48-hour infusion of P188 resulted in a 42% reduction in myocardial infarct size and a 38% improvement in left ventricular function in this dog model of 90 minutes of coronary artery occlusion and 72 hours of reperfusion.

Several studies have examined the effects of P188 in animal models of **cerebral ischaemia.** P188 was associated with improved survival in a rat model of reversible global ischaemia and with improved blood flow in a rabbit model of non-reversible focal cerebral ischaemia. Studies using non-reversible middle cerebral artery occlusion in the rat (Spontaneously Hypertensive and Wistar) and rabbit demonstrated benefits associated with P188 on infarct size, cerebral blood flow, and neurological function. In other studies of ischaemia, P188 reduced the incidence of hind-limb paralysis in rabbits when administered prior to spinal cord ischaemia, and improved survival and neurologic function in dogs following hypothermic circulatory arrest.

Toxicology

Acute Toxicology Data:

LD_{50} (Mouse IV): 1 g/kg

LD_{50} (Mouse Oral): 15 g/kg

LD_{50} (Mouse SC): 5.5 g/kg

LD_{50} (Rat IV): 7.5 g/kg

LD_{50} (Rat Oral): 9.4 g/kg

(Continued)

Chronic Toxicology Data: Single intravenous bolus administration is reported (BASF 1978) to produce no toxic effects in rabbits and dogs at dosages of up to 1.0 g/kg or 0.5 g/kg, respectively. There is some controversy over the acute LD_{50} by the intravenous route to rats and mice (Kibbe et al. 2000; BASF 1978). The value for rats would be in the range 3.95–7.5 g/kg and for mice in the range 1.0–5.5 g/kg. In 14-day repeat-dose studies by the intravenous bolus route of administration, non-toxic doses were established for dogs and rabbits up to 0.5 g/kg/day. A longer study in dogs for 50 days with bolus intravenous dosing for 5 days/week at 0.1 g/kg induced signs of local irritation at injection sites.

In addition to these studies, there has been a significant amount of data published concerning the use of P188 in intravenous dose formulations of fat emulsions (Lever and Baskys 1957; Edgren 1960; Thompson et al. 1965; Schuberth and Wretkind 1961; Thompson et al. 1963; Singleton et al. 1960; Krantz et al. 1961). Typically, these formulations have comprised up to 5% concentrations in intravenous bolus injections and up to 3% in intravenous infusions. In all of these studies, performed mostly during the 1960s, the changes seen can be attributed to the lipid content of the formulation and not to the emulsifying agent (P188) itself.

Relative to the proposed indication of use, P188 has been studied in three GLP, 28-day continuous intravenous administration studies (dogs and rats). The only consistent toxic response associated with P188 was a dose-related osmotic nephrosis, which was completely reversible. These data are reported by CytRx Corporation in a summary document on the product FLOCOR™, a highly purified form of P188. However, there are no indications of the dose levels used. It would appear that these studies were performed using the more purified form of P188 (FLOCOR) rather than the commercial grade P188 (RheothRx). It was use of the latter product in a clinical trial (Maynard et al. 1998) that reportedly resulted in acute renal dysfunction characterised by rises in circulating creatinine levels. This necessitated the suspension of the clinical trial. In view of the

(Continued)

relatively low toxicity shown by P188 in animal studies, it is worth putting this finding into perspective. The inclusion criteria for patients on this trial included ongoing symptoms of acute myocardial infarction (AMI) of at least 30 minutes duration, various abnormalities of ECG and a baseline creatinine level not exceeding 2.5 mg/dL. Consequently, these patients were suffering some degree of effect on the body fluid homeostatic mechanism of renal function. One of the criteria for determining acute renal dysfunction in this study was set at a >50% increase in creatinine to a value >2.0 mg/dL. Some of the patients may have already had fluctuating creatinine levels above this level at inclusion stage. According to work in healthy animals, the potential for P188 to cause renal dysfunction or toxicity has not been observed. However, in AMI patients already suffering circulatory stresses, P188 can apparently exacerbate secondary effects on renal function. Further to these renal changes, P188 has been shown to induce a general phospholipidosis in rats (Magnusson et al. 1986). Following daily intravenous dosing to rats over 28 days, P188 induced pulmonary foam cells at dose levels of 500 and 1000 mg/kg/day and slight focal degenerative changes in the proximal tubules of the kidneys at dosage levels of 100, 200, 500, and 1000 mg/kg/day. The cytoplasm of the pulmonary foam cells contained lipid droplets, phospholipids being the most essential constituent. A number of drugs, of differing pharmacological action, can give rise to pulmonary foam cells in rats (Reason 1981; Kehrer and Kacwe 1985) although the mechanism of the induced phospholipidosis in the rat is not wholly understood.

Poloxamer 407 (Pluronic F127)

Pharmacology

This poloxamer (P407) appears to have been less extensively studied in comparison with other pluronics in the series (such as P188).

Recent studies have indicated that when administered systemically, at high dose levels, P407 produces marked effects on blood cholesterol levels. A single intraperitoneal injection of about 1200 mg/kg to rats has been shown to produce

(Continued)

a very marked hypercholesterolaemia that persisted for at least 96 hours (Johnston and Palmer 1997). This effect of P407 was associated with a marked increase in the activity of 3-hydroxy-3-methylglutaryl coenzyme A (HMG-CoA) reductase, the rate-limiting enzyme in cholesterolgenesis. However, P407 did not inhibit HMG-CoA reductase *in vitro* and the rise in blood cholesterol level did not show a closely time-related association with increased enzyme levels in microsomal fraction from livers of P407-treated animals. It was concluded that the activity of HMG-CoA reductase was regulated by some indirect mechanism(s) after injection of P407 in rats. Later, it was demonstrated that P407 produced a dose-related hypercholesterolaemia and hypertriglyceridaemia in both mice and rats (Palmer et al. 1997). It was considered that these effects were due to both an increase in activity of HMG-CoA reductase and to an inhibition of lipoprotein lipase. Chronic administration (145 days) of P407 to C57BL/6 mice produced atherogenic lesions in the aorta to an extent similar to those seen in mice fed a high cholesterol diet for the same period. The intraperitoneal levels of P407 used in these experiments were very high, approximately 1200 mg/kg in rats and 500 mg/kg in mice.

Toxicology

Acute toxicology data:

LD_{50} (Mouse IV): 7.5 g/kg
LD_{50} (Mouse SC): 5.5 g/kg
LD_{50} (Rat SC): 6.9 g/kg
LD_{50} (Rat IV): 7.5 g/kg
LD_{50} (Rat Oral): 10.0 g/kg
LD_{50} (Rabbit Dermal): >5 g/kg

Chronic toxicology data:

When P407 was administered intravenously as a 5% solution to dogs, 68–75% of the total dose injected was recovered in urine over a 30-hour period. As mentioned previously, this poloxamer has been further developed in subcutaneous/intramuscular sustained-release formulations.

(Continued)

Other Comments   Poloxamers are non-ionic polyoxyethylene-
                  polyoxypropylene copolymers used primarily in
                  pharmaceutical formulations as emulsifying or
                  solubilising agents. The polyoxyethylene segment
                  is hydrophilic while the polyoxypropylene
                  segment is hydrophobic. All of the poloxamers are
                  chemically similar in composition, differing only
                  in the relative amounts of propylene and ethylene
                  oxides added during manufacture. Poloxamers are
                  used as emulsifying agents in intravenous fat
                  emulsions.
                  (From *Handbook of Pharmaceutical Excipients*)

	Vehicle/Excipient Data Sheet
Chemical Name	**Polyethylene Glycols**
CAS ID Number	25322-68-3
Molecular Weight	36.46
Molecular Formula	$HOCH_2(CH_2OCH_2)_mCH_2OH$, where m represents the average number of oxyethylene groups.
Solubility in Water	All grades of polyethylene glycol are soluble in water at 20°C.
pH	Most effective at pH 5–7
Osmolality	7.5% v/v PEG400 is iso-osmotic with serum.
Surface Tension	52–60 mN/m at 20°C
Viscosity	105–130 mPa.s at 20°C
Stability	Polyethylene glycols exhibit good chemical stability in air and in solution, although grades with a molecular weight less than 2000 are hygroscopic. Polyethylene glycols do not support microbial growth, and they do not become rancid. (From *Handbook of Pharmaceutical Excipients*)
Toxicity	Polyethylene glycols are widely used in a variety of pharmaceutical formulations. Generally much less toxic than ethylene glycol, they are regarded as having a very low order of toxicity and as non-irritant materials, although they are considered to be skin and eye irritant. Adverse reactions to polyethylene glycols have been reported, the greatest toxicity being with glycols of low molecular weight. However, the toxicity of glycols is relatively low. LD_{50} (Rat Oral): 33.75 g/kg (From *Handbook of Pharmaceutical Excipients*)
Other Comments	PEG300 and PEG400 are typically used as the vehicle for parenteral dosage forms.

Vehicle/Excipient Data Sheet

Chemical Name	**Polyoxyl Castor Oil, Hydrogenated Polyoxyl Castor Oil, Macrogolglycerol Ricinoleate, Macrogolglycerol Hydroxystearate, Polyoxyl 35 Castor Oil, Polyoxyl 40 Hydrogenated Castor Oil**
CAS ID Number	61791-12-6
Molecular Weight	Polyoxyethylene castor oil derivatives are complex mixtures of various hydrophobic and hydrophilic components. Members within each range have different degrees of ethoxylation (moles)/PE units as indicated by their numerical suffix (n). The chemical structures of the polyethoxylated hydrogenated castor oils are analogous to polyethoxylated castor oils with the exception that the double bond in the fatty acid chain has been saturated by hydrogenation. (From *Handbook of Pharmaceutical Excipients*)
Molecular Formula	See above
Solubility in Water	Very soluble
pH	pH 6–8 as 10% w/v in water but dependent on concentration. Often formulated with ethanol with a resultant low pH range of 2–3.
Osmolality	No data available
Surface Tension	Tends to lower surface tension of solution mixtures
Viscosity	20–40 mPa.s in a 30% aqueous solution at 25°C
Stability	Polyoxyl 35 castor oil and Polyoxyl 40 hydrogenated castor oil form stable solutions. Polyoxyethylene castor oil derivatives should be stored in a well-filled, airtight container, protected from light, in a cool, dry place. They are stable for at least 2 years if stored in the unopened original containers at room temperature (maximum 25°C). (From *Handbook of Pharmaceutical Excipients*)
Toxicity	Polyoxyl 35 castor oil (Cremophor EL): Acute toxicology data: LD_{50} (Dog IV): 0.641 g/kg LD_{50} (Mouse IV): 6.51 g/kg LD_{50} (Mouse IP): >12.5 g/kg LD_{50} (Rabbit Oral): >10 g/kg LD_{50} (Rat Oral): >6.4 g/kg 28-day intravenous studies in rats showed dose levels of 300–900 mg/kg/day to be well tolerated.
Other Comments	Polyoxyethylene castor oil derivatives are non-ionic solubilisers and emulsifying agents used in parenteral pharmaceutical formulations. The most commonly used are the polyoxyl 35 castor oil (Cremophor EL) and polyoxyl 40 hydrogenated castor oil (Cremophor RH 40). (From *Handbook of Pharmaceutical Excipients*)

Vehicle/Excipient Data Sheet

Chemical Name	**Potassium Chloride**
CAS ID Number	7447-40-7
Molecular Weight	74.55
Molecular Formula	KCl
Solubility in Water @ 20°C	1 in 2.8
pH	7.0 for a saturated aqueous solution at 15°C
Osmolality	1.19% w/v solution is iso-osmotic with serum.
Surface Tension	At 20°C, 73.46 mN/m at 5% w/v concentration and 74.29 mN/m at 10% w/v concentration
Viscosity	Approximately 890 mPa.s at 25°C
Stability	Potassium chloride is stable and should be stored in a well-closed container in a cool, dry place. (From *Handbook of Pharmaceutical Excipients*)
Toxicity	Potassium chloride is widely used, but rapid injection of strong potassium chloride solutions can cause cardiac arrest. Acute toxicology data: LD_{50} (Guinea pig Oral): 2.5 g/kg LD_{50} (Mouse IP): 1.18 g/kg LD_{50} (Mouse IV): 0.12 g/kg LD_{50} (Mouse Oral): 0.38 g/kg LD_{50} (Rat IP): 0.66 g/kg LD_{50} (Rat IV): 0.14 g/kg LD_{50} (Rat Oral): 2.6 g/kg (From *Handbook of Pharmaceutical Excipients*)
Other Comments	Potassium chloride is widely used in a variety of parenteral formulations to produce isotonic solutions.

Vehicle/Excipient Data Sheet

Chemical Name	**Propylene Glycol**
CAS ID Number	57-55-6
Molecular Weight	76.09
Molecular Formula	$C_3H_8O_2$
Solubility in Water	Miscible with water at 20°C
pH	pH 8–9 at 20°C
Osmolality	2% v/v is iso-osmotic with serum.
Surface Tension	40.1 mN/m at 25°C
Viscosity	58.1 mPa.s (58.1 cP) at 20°C
Stability	At cool temperatures, propylene glycol is stable in a well-closed container, but at high temperatures, in the open, it tends to oxidise, giving rise to products such as propiomaldehyde, lactic acid, pyruvic acid, and acetic acid. Propylene glycol is chemically stable when mixed with ethanol (95%), glycerin, or water; aqueous solutions may be sterilised by autoclaving.
	Propylene glycol is hygroscopic and should be stored in a well-closed container, protected from light, in a cool, dry place.
	(From *Handbook of Pharmaceutical Excipients*)
Toxicity	The primary pharmacological action of propylene glycol is to produce central nervous system (CNS) depression, as with ethanol. However, elimination of propylene glycol is slower that that of ethanol, and its actions are thus relatively prolonged (Klassen 1985).
	Acute toxicology data:
	LD_{50} (Mouse IP): 9.72 g/kg
	LD_{50} (Mouse IV): 6.63 g/kg
	LD_{50} (Mouse Oral): 22.0 g/kg
	LD_{50} (Mouse SC): 17.37 g/kg
	LD_{50} (Rat IM): 14.0 g/kg
	LD_{50} (Rat IP): 6.66 g/kg
	LD_{50} (Rat IV): 6.42 g/kg
	LD_{50} (Rat Oral): 20.0 g/kg
	LD_{50} (Rat SC): 22.5 g/kg
	Injections (2.1–5.2 g/kg bwt) to rabbits produced kidney damage (Kesten et al. 1939). Haemolysis has been induced in cows, dogs, rabbits, chickens, rats, and sheep given 10–80% solutions in water, saline, or ethanol (Gentry and Black 1976; Gross et al. 1979; Guy 1990; Lehman and Newman 1937; MacCannell 1969; Potter 1958; Vaille et al. 1968), with aqueous solutions being more active. Concentrations of around 2–3% in saline were ineffective

(*Continued*)

in rats (Vaille et al. 1968) and dogs (MacCannell 1969), although this treatment did produce slight changes in blood flow in the heart and kidneys of dogs. According to a brief abstract, a 40–55% solution infused continuously for 14 days (unspecified dose!) produced slight liver and kidney damage in rats. Dogs were less affected (Pickersgill et al. 1994).

Rapid injections of 0.5–1.0 g/kg bwt in cats resulted in disturbances in heart rhythm and rapid transient falls in blood pressure. Slower infusions (1 mL/min or less) of 3.6 g/kg bwt had minimal effects (Louis et al. 1967).

Chronic toxicology data:

There is quite a lot of data relating to effects of propylene glycol in various animal species following exposure by the oral route. However, by injection, doses in the order of 6.2–8.3 g/kg bwt/day given to rats intraperitoneally for 3 days changed a number of liver enzyme activities (Dean and Stock 1974). An abstract reported haemoglobin in the urine (presumably arising from red blood cell haemolysis) of rats that had received about 240 mg/kg bwt/day by intravenous injection for 30 days. No effects were seen in 10 males given 80 mg/kg bwt/day for 30 days (Horspool and Joseph 1991).

Other relevant considerations include an investigation of the metabolism of propylene glycol intravenously infused into rabbits in which a complex and unpredictable metabolic and kidney clearance was found. This led the investigators to suggest that long-term exposure to propylene glycol may result in toxicity (Yu and Sawchuk 1987). Similar marked differences in propylene glycol accumulation has been reported in humans between individuals because of variability in clearance following intravenous injection (Speth et al. 1987).

In conclusion, it would appear that following prolonged intravenous infusion of propylene glycol to rats there would be the risk of kidney damage in some individuals and red blood cell haemolysis above certain threshold exposure. Because of the potential variability in metabolism and clearance of this chemical, it is difficult to be precise about this threshold dose. This can be approached from two directions: the dose in terms of mg/kg bwt/day, or the concentration of the solution. When considering both these approaches based on the given information, two very different answers are achieved. For example, using the dose approach, it would appear that

(Continued)

a dose of 200 mg/kg bwt/day should not be exceeded to avoid potential changes. With rats being infused at a rate of approximately 1 mL/hour this would equate to a dose volume of 4 mL/kg bwt/hour with a dose concentration of 0.2%. But solutions of around 50% concentration have allegedly been infused intravenously to rats with only slight effects. Certainly, concentrations of up to 10% have been well tolerated, and as long as the delivery rate is slow the danger of haemolysis should be controlled. The potential for renal or hepatic changes will be dependent on the duration of the exposure.

Other Comments   Propylene glycol has become widely used as a solvent, extractant, and preservative in a variety of parenteral and non-parenteral pharmaceutical formulations. It is a better general solvent than glycerin and dissolves a wide variety of materials, such as corticosteroids, phenols, sulpha drugs, barbiturates, vitamins (A and D), most alkaloids, and many local anaesthetics.

(From *Handbook of Pharmaceutical Excipients*)

Vehicle/Excipient Data Sheet	
Chemical Name	**Pyrrolidone (2-Pyrrolidon),** also related **N-Methyl Pyrrolidone**
CAS ID Number	616-45-5 (Pyrrolidone)
Molecular Weight	85.11 (Pyrrolidone); 99.14 (N-Methylpyrrolidone)
Molecular Formula	C_4H_7NO (Pyrrolidone); C_5H_9NO (N-Methylpyrrolidone)
Solubility in Water	Miscible in water at 20°C
pH	8.2–10.8 for a 10% v/v aqueous solution
Osmolality	2.5% v/v is iso-osmotic with serum.
Surface Tension	No data available
Viscosity	13.3 mPa.s at 25°C
Stability	Pyrrolidone is chemically stable and, if it is kept in unopened original containers, the shelf-life is approximately one year. Pyrrolidone should be stored in a well-closed container protected from light and oxidation, at temperatures below 20°C. (From *Handbook of Pharmaceutical Excipients*)
Toxicity	Pyrrolidones are used mainly in veterinary injections, but N-methylpyrrolidone is considered a poison when injected by the intravenous route. 2-Pyrrolidone: Acute toxicology data: LD_{50} (Guinea pig Oral): 6.5 g/kg LD_{50} (Rat Oral): 6.5 g/kg N-methylpyrrolidone: Acute toxicology data: LD_{50} (Mouse IP): 3.05 g/kg LD_{50} (Mouse IV): 0.155 g/kg LD_{50} (Mouse Oral): 5.13 g/kg LD_{50} (Rabbit SC): 8.0 g/kg LD_{50} (Rat IP): 2.472 g/kg LD_{50} (Rat IV): 0.081 g/kg LD_{50} (Rat Oral): 3.914 g/kg
Other Comments	None

Vehicle/Excipient Data Sheet

Chemical Name	**Sodium Acetate (Sodium Acetate Trihydrate, Sodium Acetate Hydrate)**
CAS ID Number	Sodium acetate anhydrous 127-09-3 Sodium acetate trihydrate 6131-90-4
Molecular Weight	82.0 (anhydrous) 136.1 (trihydrate)
Molecular Formula	$C_2H_3NaO_2$ (anhydrous) $C_2H_3NaO_2.3H_2O$ (trihydrate)
Solubility in Water @ 20°C	Soluble 1 in 0.8 v/v
pH	7.5–9.0 (5% w/v aqueous solution)
Osmolality	1.19% w/v solution is iso-osmotic with serum.
Surface Tension	Dependent on concentration – no data available
Viscosity	Dependent on concentration – no data available
Stability	Sodium acetate is stable and should be stored in a well-closed container in a cool, dry place. (From *Handbook of Pharmaceutical Excipients*)
Toxicity	LD_{50} (Rat Oral): 3.53 g/kg LD_{50} (Mouse IV): 0.38 g/kg LD_{50} (Mouse SC): 8.0 g/kg (From *Handbook of Pharmaceutical Excipients*)
Other Comments	Sodium acetate is used as part of a buffer system when combined with acetic acid.

Vehicle/Excipient Data Sheet	
Chemical Name	**Sodium Chloride**
CAS ID Number	7647-14-5
Molecular Weight	58.44
Molecular Formula	NaCl
Solubility in Water @ 20°C	Soluble 1 in 2.8
pH	6.7–7.3 (saturated aqueous solution)
Osmolality	0.9% w/v solution is iso-osmotic with serum.
Surface Tension	Not applicable
Viscosity	Not applicable
Stability	Sodium chloride is stable and should be stored in a well-closed container in a cool, dry place. (From *Handbook of Pharmaceutical Excipients*)
Toxicity	In rats, the minimum lethal intravenous dose is 2.5 g/kg body weight. LD_{50} (Mouse IP): 6.61 g/kg LD_{50} (Mouse IV): 0.65 g/kg LD_{50} (Mouse Oral): 4.0 g/kg LD_{50} (Mouse SC): 3.0 g/kg LD_{50} (Rat Oral): 3.0 g/kg (From *Handbook of Pharmaceutical Excipients*)
Other Comments	Sodium chloride is widely used in a variety of parenteral formulations to produce isotonic solutions.

Vehicle/Excipient Data Sheet	
Chemical Name	**Sodium Hydroxide**
CAS ID Number	1310-73-2
Molecular Weight	40
Molecular Formula	NaOH
Solubility in Water @ 20°C	Very soluble – 1 part in less than 1 part water
pH	12 (0.05% w/w aqueous solution)
Osmolality	Not applicable
Surface Tension	Not applicable
Viscosity	Not applicable
Stability	Sodium hydroxide should be stored in an airtight non-metallic container in a cool, dry place. When exposed to air, sodium hydroxide rapidly absorbs moisture and liquefies, but subsequently becomes solid again owing to absorption of carbon dioxide and formation of sodium carbonate. (From *Handbook of Pharmaceutical Excipients*)
Toxicity	LD_{50} (Mouse IP): 0.04 g/kg LD_{50} (Rabbit Oral): 0.5 g/kg
Other Comments	Sodium hydroxide is widely used in pharmaceutical formulations to adjust the pH of solutions. (From *Handbook of Pharmaceutical Excipients*)

Vehicle/Excipient Data Sheet

Chemical Name	**Soybean Oil (Refined Soya Oil)**
CAS ID Number	8001-22-7
Molecular Weight	Not applicable
Molecular Formula	A typical analysis of refined soybean oil indicates the composition of the acids, present as glycerides, to be linoleic acid (50–57%); linolenic acid (5–10%); oleic acid (17–26%); palmitic acid (9–13%); stearic acid (3–6%). Other acids are present in trace quantities. (From *Handbook of Pharmaceutical Excipients*)
Solubility in Water	Practically insoluble in water at 20°C
pH	Not applicable
Osmolality	No data available
Surface Tension	25 mN/m (25 dynes/cm) at 20°C
Viscosity	50.09 mPa.s (50.09 cP) at 25°C
Stability	Soybean oil is a stable material if protected from atmospheric oxygen. (From *Handbook of Pharmaceutical Excipients*)
Toxicity	LD_{50} (Mouse IV): 22.1 g/kg LD_{50} (Rat IV): 16.5 g/kg (From *Handbook of Pharmaceutical Excipients*)
Other Comments	Soybean oil is widely used intramuscularly as a drug vehicle or as a component of emulsions used in parenteral nutrition regimens. (From *Handbook of Pharmaceutical Excipients*)

Vehicle/Excipient Data Sheet	
Chemical Name	**Sulphobutylether β-Cyclodextrin**
CAS ID Number	1824100-00-0
Molecular Weight	2163
Molecular Formula	$C_{42}H_{70-n} O_{35} \cdot (C_4H_8SO_3Na)_n$ (where n = approximately 6.5)
Solubility in Water	Greater than 1 in 2 parts of water at 25°C
	1.85 g/100mL at 25°C (Loftsson and Brewster 1996)
pH	4.0–6.8 (30% w/v aqueous solution)
Osmolality	9.5–11.4% w/v is iso-osmotic with serum
	(From *Handbook of Pharmaceutical Excipients*)
Surface Tension	No data available
Viscosity	1.75 mPa.s for a 8.5% w/w aqueous solution at 25°C
Stability	Store in well-closed containers. Sulphobutylether-β-cyclodextrin solutions may be autoclaved.
	(From *Handbook of Pharmaceutical Excipients*)
Toxicity	Reversible kidney vacuolation (see hydroxypropyl-beta-cyclodextrin)
	LD_{50} (Mouse IP): >10 g/kg (Rajewski et al. 1995)
Other Comments	Sulphobutylether-β-cyclodextrin is widely used parenterally generally at concentration <30% w/v.

References

Babbitt D, Forman M, Jones R, Bajaj A and Hoover R. 1990. Prevention of neutrophil-mediated injury to endothelial cells by perflurochemicals. *Am. J. Pathol.* 136: 451–459.

Bajaj A, Cobb M, Virmani R, Gay J, Light R and Forman R. 1989. Limitation of myocardial reperfusion injury by intravenous perflurochemicals. *Circulation* 79: 645–656.

BASF Wyandotte Corporation. 1978. *Pluronic Polyols: Toxicity and Irritation Data.* Form no. 3012–767, pp. 1–31. Wyandotte, Michigan.

Benner K, Gaehtgens P and Frede K. 1973. Aggregation of human red blood cells (RBC). *Bibl. Anat.* 12: 208–215.

Bond GR et al. 1989. Dimethyl sulfoxide-induced encephalopathy. *Lancet I*: 1134–1135.

Brayton CF. 1986. Dimethyl Sulfoxide (DMSO): A review. *Cornell Vet.* 76(1): 61–90.

Brewster ME, Estes KS and Bodor N. 1990. An intravenous toxicity study of 2-HP-β-CD: A useful drug solubiliser, in rats and monkeys. *Int. J. Pharm.* 59: 231–243.

Brobyn RD. 1975. The human toxicology of dimethyl sulfoxide. *Am. NY Acad. Sci.* 243: 497–506.

Coussement W, van Cauteren H, Vandenberghe J, Vanparys P, Teuns G, Lampo A and Marsboom R. 1990. Toxicological profile of HP-β-CD in laboratory animals. In: *Minutes of the 5th International Symposium on Cyclodextrins*, Duchene D (ed.), pp. 522–524. Editions de Sante.

Dean ME and Stock BH. 1974. Propylene glycol as a drug solvent in the study of hepatic microsomal enzyme metabolism in the rat. *Toxic. Appl. Pharmac.* 28: 44.

Edgren B. 1960. Tolerance of the dog to intravenous fat emulsion. *Acta Pharmacol. et Toxicol.* 16: 260–269.

Forman MB, Pitarys CJ, Vildibill HD, Lambert TL, Ingram DA, Virmani R and Murray JJ. 1992. Pharmacologic perturbation of neutrophils by Fluosol results in a sustained reduction in infarct size in the canine model of reperfusion *J. Am. Coll. Cardiol.* 19: 205–216.

Fowler JE. 1981. Prospective study of intravesical dimethyl sulfoxide in the treatment of suspected early interstitial cystitis. *Urology* 18: 21–26.

Gentry PA and Black WD. 1976. Influence of pentobarbital sodium anaesthesia on haematologic values in the dog. *Am. J. Vet. Res.* 37: 1349.

Gross DR, Kittman JV and Adams HR. 1979. Cardiovascular effects of intravenous administration of propylene glycol and of oxytetracycline in propylene glycol in calves. *Am. J. Vet. Res.* 40: 783–791.

Grover FL, Heron MW, Newman MM and Paton BC. 1969. Effect of non-ionic surface-active agent on blood viscosity and platelet adhesiveness. *Circulation* (Suppl. 1) 39 & 40: 249–252.

Grover FL, Kahn RS, Heron MW and Paton BC. 1973. In: Amphiphilic block copolymers: Self-assembly and applications. Alexandridis P and Lindman B (Eds.). *Arch. Surg.* 106: 307–310.

Grover FL, Amundsen D, Warden JL, Fosburg RG and Paton BC. The effect of pluronic F-68 on circulatory dynamics and renal and carotid artery flow during haemorrhagic shock. 1974. *J. Surg. Res.* 17: 30–35.

Guy RC. 1990. *J. Am. Coll. Toxicol.* B1: 53.

Haschek WM et al. 1989. Effects of dimethyl sulfoxide (DMSO) on pulmonary fibrosis in rats and mice. *Toxicology* 54(2): 197–205.

Horspool F and Joseph EC. 1991. *Hum. Exp. Toxicol.* 10: 507 (Abstract).

Hymes AC, Safavian MH, Arbula A and Baute P. 1968. *J. Thorac. Cardiovasc. Surg.* 15: 16.

Hymes AC, Safavian MH and Gunther T. 1971. The influence of an industrial surfactant Pluronic F-68 in the treatment of haemorrhagic shock. *J. Surg. Res.* 11: 191–197.

Hunter R, Bennet B and Check I. 1990. The effect of poloxamer 188 on the rate of *in vitro* thrombolysis mediated by t-PA and streptokinase. *Fibrinolysis* 4: 117–123.

Johnston TP and Palmer WK. 1997. Effect of poloxamer 407 on the activity of microsomal 3-hydroxy-3-methylglutaryl CoA reductase in rats. *J. Cardiovasc. Pharmacol.* 29: 580–585.

Justicz AG, Farnsworth WV, Sobermann MS, Tuvlin MB et al. 1991. Reduction of myocardial infarct size by P188 and mannitol in a canine model. *Am. Heart J.* 122: 671–680.

Karaca M, Bilgin UY, Akar M et al. 1991. Dimethyl sulphoxide lowers ICP after closed head trauma *Eur. J. Clin. Pharmacol.* 40(1): 113–114.

Katakam M, Ravis WR and Banga AK. 1997. Controlled release of human growth hormone in rats following parenteral administration of poloxamer gels. *J. Controlled Release* 49: 21–26.

Kehrer JP and Kacwe S. 1985. Systematically applied chemicals that damage lung tissue. *Toxicology* 35: 251–293.

Kesten HD et al. 1939. *Archs. Path.* 27: 447.

Kibbe AH. 2000. In: *Handbook of Pharmaceutical Excipients*, 3rd edition, Kibbe AH (ed.), pp. 386–388. London: The Pharmaceutical Press; Washington: American Pharmaceutical Association.

Klassen CD. 1985. In: *The Pharmacological Basis of Therapeutics*. Goodman LS and Gillman AG (eds). Macmillan.

Kocsis JJ et al. 1975. In: Biological effects of the metabolites of dimethyl sulfoxide. *Ann. NY Acad. Sci.* 243: 104–109.

Krantz JC, Cascorbi HF, Helrich M, Burgison RM, Gold MI and Rudo F. 1961. A note on the intravenous use of anaesthetic emulsions in animals and man with special reference to methoxyfluane. *Anaesthesiology* 22: 491–492.

Lawrence HJ and Goodnight SH. 1983. Dimethyl sulfoxide and extravasation of anthracycline agents (letter). *Ann. Intern. Med.* 98: 1025.

Lechmann TH and Reinhart WH. 1998. The non-ionic surfactant poloxamer 188 (RheothRx®) increases plasma and whole blood viscosity. *Clinical Hemorheology and Microcirculation* 18: 31–36.

Lehman AJ and Newman HW. 1937. Propylene glycol rate of metabolism, absorption and excretion with a method for estimation in body fluids. *J. Pharmac. Exp. Ther.* 60: 312.

Lever WF and Baskys B. 1957. Effects of intravenous administration of fat emulsions and their emulsifying agents. I. Effects on clearing factor activity, electrophoretic pattern and clotting time of blood in dogs. *J. Invest. Dermatol.* 28: 317–320.

Loftsson T and Brewster ME. 1996. Pharmaceutical applications of cyclodextrins. 1. Drug solubilisation and stabilisation. *J. Pharma. Sci.* 85(10): 1017–1025.

Louis S et al. 1967. *Am. Heart J.* 74: 523.

Lowe KC, Furmidge BA and Thomas S. 1995. Haemolytic properties of plutonic surfactants and effects of purification. *Art. Cells Blood Subs. & Immob. Biotech.* 23: 135–139.

MacCannell K. 1969. Hemodynamic responses to glycols and to hemolysis. *Can. J. Physiol. Pharmac.* 47(6): 563–569.

Magnusson G, Olsson T and Nyberg J-A. 1986. Toxicity of pluronic F-68. *Toxicology Letters* 30(3): 203–207.

Manja KS, Nambinarayanan TK, Krishnan B, Srinivasa Rao A and Venigopala Rao A. 1974. Effects of pluronic F-68 on surface tension and viscosity of blood and plasma. *Curr. Sci.* 43: 510–511.

Marshall LF et al. 1984. The outcome of severe closed head injury. *Neurosurgery* 14: 659–663.

Maynard C, Swenson R, Paris JA, Martin JS, Hallstrom AP, Cerqueira MD and Weaver WD. 1998. Randomised controlled trial of RheothRx (poloxamer 188) in patients with suspected acute myocardial infarction. RheothRx in myocardial infarction study group. *Am. Heart J.* 135: 797–804.

Montaguti P et al. 1994. Reproductive toxicology of the new antitussive moguisteine. *Arzneimittelforschung* 44(4): 566–570.

Muther RS and Bennett WM. 1980. Effects of dimethyl sulfoxide on renal function in man. *JAMA* 244: 2081–2083.

Nalbandian RM, Hymes AC and Henry R. 1969. *Bull. Pathol.* 10: 90.

Noel PRB et al. 1975. The toxicity of dimethyl sulfoxide (DMSO) for the dog, pig, rat and rabbit. *Toxicology* 3(2): 143–169.

Olver IN, Aisner J, Hament A et al. 1988. A prospective study of topical dimethyl sulfoxide for treating anthracycline extravasation. *J. Clin. Oncol.* 6(11): 1732–1735.

Palmer WK, Emeson EE and Johnston TP. 1997. The poloxamer 407-induced hyperlipidemic atherogenic animal model. *Med. Sci. Sports Exerc.* 29: 1416–1421.

Perez-Marrero R et al. 1988. A controlled study of dimethyl sulfoxide in interstitial cystitis. *J. Urol.* (Baltimore) 140: 36–39.

Pickersgill N et al. 1994. *Toxicology Letters,* 74 (Suppl. 1): 66 (Abstract).

Potter BJ. 1958. Haemoglobinuria caused by propylene glycol in sheep. *Br. J. Pharmac.* 13: 385.

Rajewski RA, Traiger G, Brennahan J, Jaberaboansari P, Stella VJ and Thompson DO. 1995. Preliminary safety evaluation of parenterally administered sulfoalkyl ether beta-cyclodextrin derivatives. *J. Pharm. Sci.* 84(8): 927–932.

Rayburn JR et al. 1991. *J. Appl. Toxicol.* 11(4): 253–260.

Reason MJ. 1981. *Toxicology* 20: 1–33

Robinson KA, Hunter RL, Stack JE, Hearn JA, Apkarian RP and Reubin GS. 1990. *J. Invas. Cardiol.* 2: 9–20.

Ryan PG and Wallace DMA. 1990. *J. Clin. Pharm. Ther.* 15: 1–12.

Salauze D and Cave D. 1995. Choice of vehicle for three-month continuous intravenous toxicology studies in the rat: 0.9% saline versus 5% glucose. *Laboratory Animals* 29: 432–437.

Schaer GL, Karas SP, Santoian EC, Gold C, Visner MS and Virmani R. 1990. Reduction in reperfusion injury by blood-free reperfusion after experimental myocardial infarction. *J. Am. Coll. Cardiol.* 15: 1385–1393.

Schaer GL, Hursey TL, Abrahams SL, Buddemeier K et al. 1994. Reduction in reperfusion-induced myocardial necrosis in dogs by RheothRx (poloxamer 188 N.F.) a haemorheologic agent that alters neutrophil function. *Circulation* 90: 2964–2975.

Schuberth O and Wretlind A. 1961. Intravenous infusion of fat emulsions, phosphatides and emulsifying agents. *Acto Chir. Scand.* 278: 1–21.

Simpson PJ, Fantone JC, Mickelson JK, Gallagher KP and Kucchesi BR. 1988. Identification of a time window for therapy to reduce experimental canine myocardial injury suppression of neutrophil activation during 72 hours of reperfusion. *Circ. Res.* 63: 1070–1079.

Singleton WS, Benerito RR, Talluto KF, Brown ML, DiTrapani LL and White JL. 1960. Comparative effects of oral and intravenous administration of fat emulsion on sera fractions of dogs. *Metabolism* 9: 956–965.

Speth PAJ et al. 1987. Propylene glycol pharmacokinetics and effects after intravenous infusion in humans. *Ther. Drug Monit.* 9: 255.

Thompson SW, Hartwig QL, Atik M, Fox MA and Cohn I. 1963. Some long-term effects following daily infusion of intravenous fat emulsions into dogs. *Tox. Appl. Pharmacol.* 5: 306–318.

Thompson SW, Jones LD, Ferrell JF, Hunt RD et al. 1965. *Am. J. Clin. Nutrition* 16: 43–61.

Tokioka H, Miyazaki A, Fung P et al. 1987. Effects of intracoronary infusion of arterial blood or Fluosol-DA 20% on regional myocardial metabolism and function during brief coronary artery occlusions. *Circulation* 75: 473–481.

Trice JM and Pinals RS. 1985. Dimethyl sulfoxide: A review of its use in the rheumatic disorders. *Semin. Arthritis Rheum.* 15(1): 45–60.

Vaille Ch, Debray C, Souchard M et al. 1968. Hemolytic action of propylene glycol in rats. Partial protection by sorbitol. *Annls. Pharm. Fr.* 26(1): 17–23.

Vercellotti GM, Hammerschmidt DE, Craddock PR and Jacob HS. 1982. Activation of plasma complement by perfluorocarbon artificial blood:probable mechanism of adverse pulmonary reactions in treated patients and rationale for corticosteroid prophylaxis. *Blood* 59: 1299.

Willhite CC and Katz PI. 1984. Dimethyl sulfoxide. *J. Appl. Toxicol.* 4: 155–160.

Wood DC and Wood J. 1975. Pharmacologic and biochemical considerations of dimethyl sulfoxide. *Ann. NY Acad. Sci.* 243: 7–19.

Yellowlees P et al. 1980. Dimethyl sulfoxide-induced toxicity. *Lancet* ii: 1004–1006.

Yoshikawa T. 1985. Toxicity of polycyclic aromatic hydrocarbons. I. Effect of phenanthrene, pyrene and their ozonised products on blood chemistry in rats. *Toxicol. Appl. Pharmacol.* 79(2): 218–226.

Yu DK and Sawchuk RJ. 1987. Pharmacokinetics of propylene glycol in the rabbit. *J. Pharmacokinet. Biopharm.* 15: 453.

Index

A

Abraxane®, 92
Absorption, distribution, metabolism and excretion (ADME) studies, 75, 77
Acetate buffer, 84
Acetic acid test substance data sheet, 140
Acetophenone permeability into polyethylene, 124–125
Acidaemia, definition, 42
Acid-base balance, 36–49; see also pH
  infusion forces and, 58
  map of, 46
  strong ion difference (SID), 46–47
Acidosis, 42–49
ADH, see Antidiuretic hormone (ADH)
ADME (absorption, distribution, metabolism and excretion) studies, 75, 77
Administration volumes by species, 76
Adsorption onto container walls, 121
  factors affecting, 124
Albumin buffering action, 42
Albumin synthesis affected by cosolvents, 99
Alkalaemia, definition, 42
Alkalosis, 42–49
Alveolar ventilation, 49
Alzet®, 122
Amino acid terminal amino groups buffering action, 41
Amiodarone, 100, 105
Ammonium buffer, 84
Animal body heat may affect stability, 120
Animal feed factors, 14–15
Animal welfare considerations, 66
Anion effects, 48
Anion gap, 47–48
  infusion forces and, 57–58

shift as indicative of acid-base disturbance, 48
  by species, 47
  values by species, 47
Anions, 'measured' and 'unmeasured,' 48
Antidiuretic hormone (ADH), 13
  suppression of, 14
Anti-malarials and cyclodextrins, 96
Antioxidants to improve stability, 110
Arterial blood gas
  measurement, 43
  values by species, 44
Arterioles in hydrostatic pressure transfer, 9
Ascorbate buffer, 84
Ascorbic acid as antioxidant, 110
Automated clinical chemistry analysers, 47
Avogadro's law, 3

B

Barbiturate action affected by glycofurol, 99
Benzaldehyde permeability into polyethylene, 124–125
Benzoate buffer, 84
Benzoic acid permeability into polyethylene, 124–125
Bicarbonate buffer system, 84
  in acid-base disturbance, 48
Bicarbonate ion (HCO_3^-)
  anion gap by species, 47
  plasma levels by species, 8
Bicarbonate-carbonic acid buffer system, 39–41
Bile salts for mixed micelles, 88
Binding effects, 94, 96
Biochemical profile for acid-base status, 48

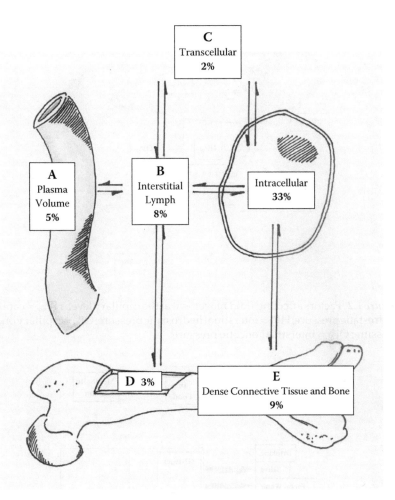

Figure 1.1 Compartments of total body water expressed as percentage of body weight. A+B+C+D+E = 27% Body Weight (Total extracellular fluid volume); A+B+C+D = 18% Body Weight (Rapidly equilibrating space).

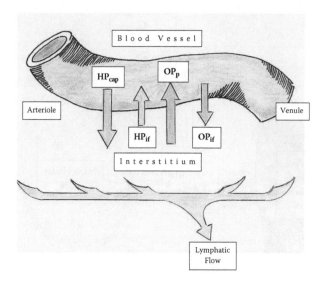

Figure 1.2 Factors affecting fluid movement at the capillary level. HP_{cap} = capillary hydrostatic pressure; HP_{if} = interstitial hydrostatic pressure; OP_p = capillary oncotic pressure; OP_{if} = interstitial oncotic pressure.

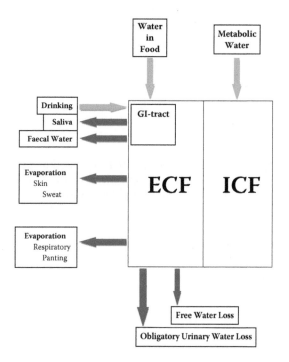

Figure 1.3 Total body water. Daily input and obligatory losses.

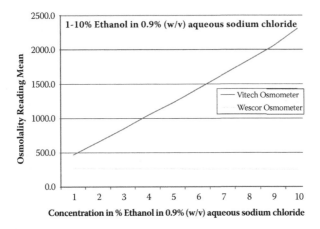

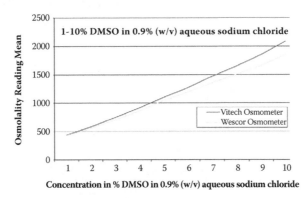

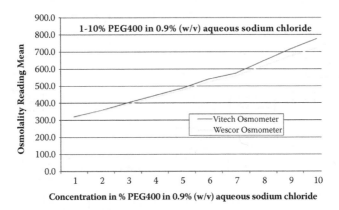

Figure 2.1 Mean osmolality reading for 1–10% DMSO, ethanol, and PEG 400 in 0.9% w/v aqueous sodium chloride measured by vapour pressure/freezing point depression.

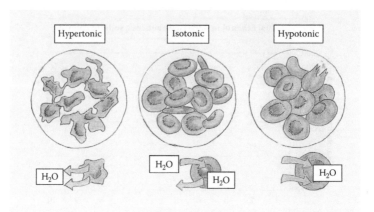

Figure 2.2 Effects of changes in tonicity on a solution of red blood cells.

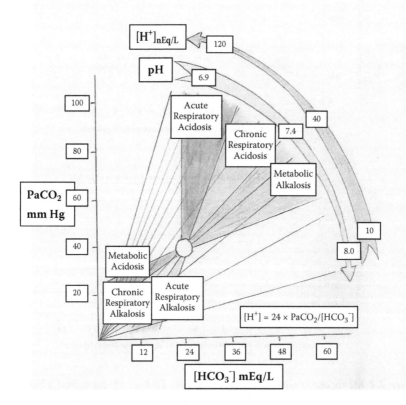

Figure 2.5 Acid-base map.

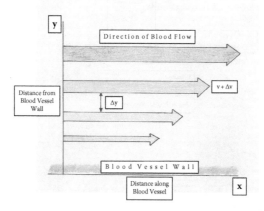

Figure 2.6 Blood flow within a blood vessel.

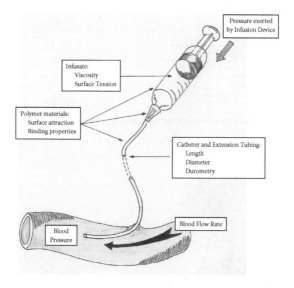

Figure 2.8 Forces affecting surface tension in an infusion system.

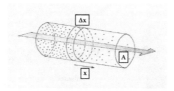

Figure 2.9 Diffusion due to a concentration gradient.

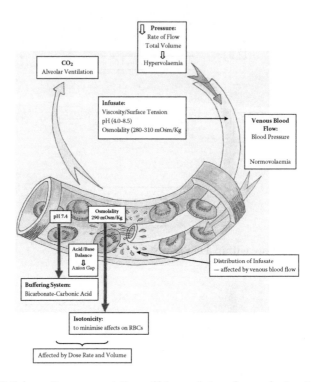

Figure 2.10 Schematic representation of bioregulatory forces during intravenous infusion delivery.

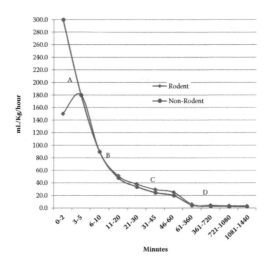

Figure 3.1 Rate of delivery versus infusion duration for rodents and non-rodents.

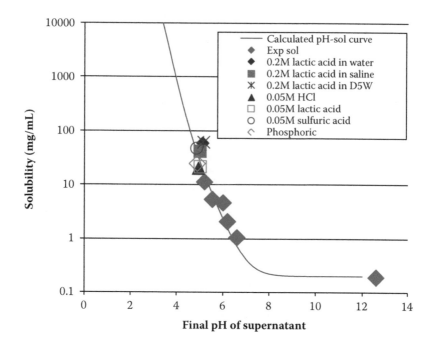

Figure 4.3 Predicted and measured solubilities for a basic compound.

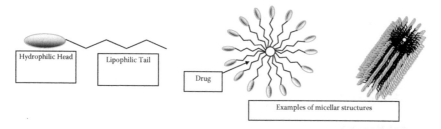

Figure 4.7 Schematic structure of surfactant molecules and types of micelle that may be formed.

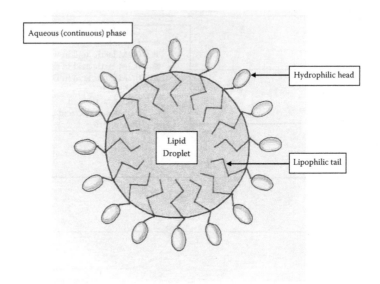

Figure 4.9 Schematic of an emulsion droplet.

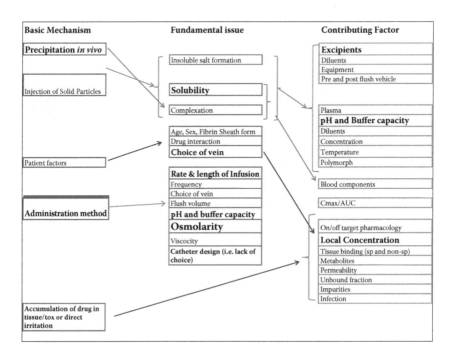

Figure 4.10 Map of injection site reaction root causes.

T - #0376 - 101024 - C8 - 234/156/12 - PB - 9781138382169 - Gloss Lamination